IT ARCHITECT SERIES

Designing
RISK in IT
INFRASTRUCTURE

DAEMON BEHR

ISBN: 978-0-9990929-0-3 (sc)
ISBN: 978-0-9966477-9-3 (e)

Lulu Publishing Services rev. date: 08/08/2018

Contents

Dedication

This book has been a labor of love and an idea that has taken several years to realize. I could not have done it without continually being pushed forward and being put back on course by my wonderfully supportive wife Tianna.

Thank you to all the editors and reviewers that have made this possible. Your hard work and insight has allowed me to portray the information provided here in a clear (and sometimes clever) way.

Thanks to the subject matter experts and industry pundits that were willing to take the time to answer my shotgun blasts of questions that often came from left field. Or what I like to call "shock and awe research".

Special thanks go to John Arrasjid and Mark Gabryjelski who have created the IT Architect Series and provided a venue for aspiring authors to go from an idea on pen and paper to ultimately a book on shelves and distribution of their works online in eBook format. The community is wonderful and through this process many great connections were made.

Also, thank you to Mark Burgess for writing the foreword and taking the time to convey the concept of the book in such an eloquent manner.

Foreword

The subject of risk and system reliability in IT is something I've dabbled with from time to time since taking up the challenge to approach IT systems from an analytical point of view in the mid 1990s. Risk assessment and reliability studies were first brought into modern engineering through the component manufacturing industry. They were based on the statistical variability inherent in mass production and were principally about estimating the likely lifetime of components in large batches. For example, when would a lightbulb likely fail, and need replacement? The failure of a lightbulb is not a high-risk occurrence, but as commodity systems became the norm, those were the same techniques that were applied to aeronautics and other mission critical systems, where human lives were weighed in the balance. The nuclear industry took on the mantle of extending risk to attempt the prediction of faults in highly explosive systems, and the notion of "root cause analysis" was born to diagnose faults in nuclear power stations.

Although an innocent name, and a convenient narrative for insurance companies, the idea of simple causation and single points of failure came to be iconic, if oversimplified, and what began as science, was turned into a politically motivated excuse to dodge responsibility or to support marketing claims. Since then, risk and causation have never really recovered from the veil of mysticism that they were intentionally thrown under by insurance companies and political interests.

Today the responsibility for design of IT software systems is shifting from engineering to business, from Op(eration)s to Dev(elopment), but the responsibility for handling risk itself remains quite homeless. It is even missing from educational narratives of our teaching institutions. It has been left to the Amazons, the Googles and the Netflix's of the world to incept a culture for risk management onto software engineers. For some, risk means only purchasing a backup and a security solution to protect against intentional harm. But harm may be intended or unintended, and the boundary between the two is blurred by system non-linearity, or what we call "complexity" in the populist terms. Many business owners believe that they only need strong encryption, and the power to fire workers for their mistakes to be protected. Major banks have told me that they are more interested in "liability" (who to sue) than in resilience and failure avoidance, because that is their cultural norm. Business problems are solved by litigation, not by engineering. Well, that was the norm before the information age. Today that thinking creates weaknesses, a legacy to be overcome.

Daemon Behr's book presents the breadth of risk analysis in a readable form for just about any reader to grasp. It is welcome reading, because it reviews both the tools and the ideas for coming to grips with the issues of risk and reliability. Whether the threats to a system are intentional (hackers, saboteurs) or unintentional (force major or so-called acts of God), it is the responsibility of system designers at all levels to equip systems for resilience, i.e. business continuity.

As my own research in promise theory has shown, the old statistical models of reliability from commodity manufacturing need to be extended to encompass network resilience if we are to understand modern systems. Our world is interconnected in an unprecedented way, and the dependencies we rest our hopes on, often unwittingly, still lead us astray. The enormous advances in safety of commercial air travel are an inspiration to all system designers, but the simple awareness of key issues has not yet propagated into the general consciousness, let alone universities. It's time that was repaired, and I hope that Daemon Behr's fine book will help to open the eyes of business leaders as well as educators to the challenges they have inherited.

Mark Burgess, Oslo March 2017

Preface

IT Architect Series

The world of infrastructure continues to evolve. As it does, there are both aspiring and experienced individuals developing, deploying, and supporting traditional and new types of infrastructures. We have evolving virtual and cloud solutions that continue to mature, while introducing new challenges and new players in this area.

Infrastructure design involves much more than technology. Just because something is built and running does not mean it is easy to use, or is adaptable to evolving business needs. IT Architects must deal with multiple vendors, technologies, skills, people, and process in a way to seamlessly support the needs of a business.

John Yani Arrasjid (VCDX-001) and Mark Gabryjelski (VCDX-23) took the journey and achieved one of the most recognized certifications in this space. They talked to publishers but could not find one that recognized the value of a series of books for IT Architects, so they created their own.

In addition, they provided a more author friendly environment that provided greater benefits to the authors.

The first book in the series, Foundation in the Art of Infrastructure Design, was the start of this series, formed under IT Architect Resource, LLC. The goal was to provide a missing source of information and reference for both established architects, and those looking to become an architect.

Most of the books in the series include the three stages of design. Conceptual models are developed during the initial gathering of input, and with design discussions to develop the customer's perspective and vision. Logical designs lay out technical and operational capabilities and a framework to select products and configurations. The physical model and design lay out the details of the technology, configurations, and operations. These three stages allow for interactive dialogue with the business teams to reach consensus on a solution supporting the needs of today and the future.

The latest books in the series, including the one you are reading now, provide details in specific domains that we believe provide additional value. In each book we include stories, examples, recommendations, and exercises. The series of books support the training and are a reference source on your journey of as an IT architect.

We encourage feedback on what you would like to see added to the series. For a full list of series books, to share your feedback, and to join the growing list of authors, please see www.itaseries.com. We look forward to hearing from you!

Authors Note

The first book "Foundation in the Art of Infrastructure Design", is, as its name implies, the foundational body of knowledge that seasoned and aspiring IT Architects should keep in their repertoire. It can be read and re-read multiple times with different desired outcomes. You can also review chapters as you progress in your career and see how you have grown and evolved in relation to the objectives of the book.

This is the second installment in the IT Architect series.

There are several outcomes from this book including, but not limited to:

- To expand the view of the IT Architect beyond technology.
- To learn how to show the dependencies and interrelations between components.
- To understand how to assess areas of risk, define criticality and exposure.
- To understand the effect of design decisions on component relationships and how to determine the costs associated with those decisions.
- To learn how to plan for multiple possible futures by looking at past and present technological and organizational evolution.
- To learn how to make the best decisions by gathering information from multiple sources and assessing probability and cost.
- To educate the reader on the design considerations for an IT infrastructure.
- To support an IT Architecture design educational course curriculum

This book complements the other books in the ITA Series, which provide a path for those wanting to pursue a career in IT Architecture.

The structure of this book is such that it is also aptly transferrable for use in a college level course. Each chapter has summary notes and review questions that are meant to highlight the main takeaways from each section within a chapter. There is no "answer key" for these review questions, as they are meant to stimulate the thought process for the concepts addressed. It should be noted here that there are many avenues to address a given problem; therefore, there is not always a single solution. When used as material in a course, the review questions can be used to stimulate discussion amongst students and delve deeper into certain areas as required.

This book is not meant to define a single path of execution for solving problems, but rather, an approach that allows for greater introspection and analysis. There are many ways to create solutions to problems, such as prescriptive workflows that guide individuals through the process of following "if this, then that" flowcharts. I do not wish to disparage these methods, as they are essential tools and of great value. However, I believe that it is fundamental to properly frame a problem before attempting to concoct a solution. Sometimes, by asking questions and getting a greater understanding of directives and drivers, you may determine the original problem is not the best one to address. Perhaps the problem itself is not fully understood by all parties involved. When this latter situation happens, there is the chance that solutions may be misaligned, or non-optimal. This generates wasted time, effort, resources and money.

This book uses a vendor agnostic approach to address concepts and technologies, and as such, it can be implemented by many different product vendors. The book focuses on perception and analysis, which does not require any costs other than time and effort, however, some vendors do make solutions that align to the precepts that are defined in this book. Those vendors will end up saving time using software to reduce the manual work required and increase efficiency.

Who Should Read This Book

This book is purposely written to be accessible to a wide audience. The goal was to demystify complex topics and distill them into something that can be easily consumed. Though it is geared primarily towards IT Infrastructure Architects, those that would gain substantial benefit from reading this book are as follows:

- Aspiring and experienced IT Infrastructure Architects.
- System Administrators and Operational Engineers
- CIOs, IT Directors and senior IT management
- Project Managers
- Technical Sales Staff
- Computer Science Engineering faculty and students
- DevOps Engineers
- Site Reliability Engineers
- Governance, Risk and Compliance (GRC) Professionals
- Information Security Professionals
- Penetration Testers

Goals and Methods

The primary goal of this book is to provide IT Architects with tools and methods for organizational introspection. This includes all technologies, all associated mushy machines (people) and all factors that influence infrastructure design, whether directly or indirectly. The book follows this flow:

Define> frame> explain> explore> analyze> introspect> understand> action.

In a more detailed fashion, we follow this method to first frame the concepts, then provide examples of what can be achieved with limitless resources. Once the reader understands that there are no limitless resources in real life, then trade-offs and decisions need to be made. These decisions need to be backed by logic and reasoning, which is then provided by means of careful analysis. We then delve into the character attributes of leaders and how risk is perceived by them. We see what actions leaders make and how those actions create ripple effects throughout the company. Introspection and optimization could be performed at every level to ensure that business objectives are being met by IT in the most efficient manner.

For those that are heavily constrained by limited resources, and a lack of (or negative) budget, we have a section that addresses those concerns. From here, we do a deep dive into the concepts of probability and costing with easily understandable mathematics and statistical analysis.

Next, we tackle the considerations of humans in the infrastructure equation. We see how to plan for the inevitable variability of illogical chaos by understanding patterns in human behavior.

In order to determine the best possible design to fit within the requirements and constraints, multiple models are designed with their risks analyzed. We review several methods for determining areas of focus, in terms of priority of alignment with business directives, cost and effort.

Towards the tail of the book, we look at induced chaos and how organizations can make use of this concept to find weaknesses and risks that would not otherwise be considered or found. Finally, we then look at what comes next in terms of certification and career advancement as the path of understanding risk is followed.

How to Use This Book

This book was written in an order that was meant to guide readers through a journey of angles of thought and discovery. Each chapter layers on the last to provide a greater insight to the material presented. However, depending on what the desired outcomes for the reader are, it can also be read in a non-linear manner. I have thus tagged the chapters with one or more characteristics. Depending on the desired outcomes and focus, readers can review chapters based on their need.

- Technical (TECH)
- Theoretical (THEO)
- Practical (PRAC)

Below is a summary of the intent of each chapter, along with the corresponding characteristics.

Chapter 1 – Overview

This chapter provides a background to the core concepts of risk and lays out the objectives for the rest of the book. (THEO)

Chapter 2 – Indestructible

This chapter dives deep into many technical topics that one should be aware of when building highly scalable, available, reliable and recoverable infrastructures. (TECH), (THEO)

Chapter 3 – Know What You Don't Know

This chapter addresses topics that are not often discussed by technical staff, nor are they brought up by most business units. They are intangible, but directly affect many aspects of design. (THEO)

Chapter 4 – MacGyver's and Gamblers

This chapter brings light to risks associated with personality, leadership and management. It also discusses methodologies and initiatives that can greatly advance the IT maturity of an organization. (THEO), (PRAC)

Chapter 5 – Guerilla IT for the SMB

This chapter is about making things work with extremely limited resources and the concepts that should be considered. (TECH), (PRAC)

Chapter 6 – Math is the Language of Quantitative Risk (Or how I learned to stop worrying and love logic)

This chapter focuses on mathematical concepts that are used by several industries and sciences in a foundational way. They are, however, discussed from an accessible and directly practical manner. (THEO), (PRAC)

Chapter 7 – The Human Element

This chapter has more psychology in it than you will find in many technical books. It is all about how people make decisions, their motivations and how to understand and predict outcomes whether they are logical or not. (THEO)

Chapter 8 – Modeling Outcomes

This chapter is about looking to the future and gathering information from many sources to make design decisions that have immediate and long-term consequences. (THEO), (PRAC)

Chapter 9 – Visual Models

This chapter provides ways of visually representing risk, dependencies and associations. It provides a means of brainstorming for easy wins as well as a framework for continual improvement and action. (TECH), (PRAC)

Chapter 10 – Red Teams and Robots

This chapter discusses several means of proactively attacking your own infrastructure through a controlled means to identify areas that need to be addressed. This is also a means of continual improvement and optimization. (THEO), (TECH), (PRAC)

Chapter 11 – Frameworks and Certifications

This chapter discusses several risk analysis frameworks and risk related certifications. (THEO), (PRAC)

Appendix A – Glossary of Terms

Terms and acronyms defined with references.

About the Author

Daemon Behr is a Solutions Architect at Scalar Decisions. His current role is Technical Account Manager and Senior Architect for large enterprise clients. He has over 20 years of industry experience creating solutions for Service Providers, Power and Utility companies and Financial Services Organizations.

Daemon has specialized in providing virtualization and converged / hyperconverged infrastructure solutions. Previously, Daemon was an Architect for one of the largest Telecommunications Service Providers in Canada. He has taught courses on infrastructure design and security at the British Columbia Institute of Technology and the University of British Columbia.

Daemon has presented talks at VMWorld, VMUG UserCon, USENIX LISA, the OpenStack Summit, vBrownBag and other events across Canada and the US.

Acknowledgments

I would like to thank the following people for helping develop and review the material included herein. Thank you to my family, friends and peers who supported my efforts in creating this work.

Thank you to the following. They have contributed to the creation of this book with their ideas, feedback, and critical insight. John Arrasjid, Mark Gabryjelski, Sandra Mereos Crosswell, Roger Singh.

Thank you to reviewers including those individuals above and James Bowling, Jonathan Copeland, Brandon Hahn, Michael A. Haines (Fortinet, Inc.), Angelo Luciani, Christian Mohn (Proact IT Norge AS), Manish Patel, Melissa Palmer, Iwan Rahabok, Gregg Robertson, Benjamin Troch, Steven Bochinski, Adnan Mahmood, Rene Van Den Bedem.

- Daemon Behr

Share your feedback!

As the reader of this book, you are our most important critic and commentator. We value your opinion and want to know what we're doing right, what we could do better, in what other areas you'd like to see us publish in, and any other words of wisdom you're willing to pass our way.

We welcome your comments. You can reach us to let us know what you did or didn't like about this book as well as what we can do to make this book better.

When you write, please be sure to include this book's title, as well as your name, e-mail address, and phone number. We will carefully review your comments, but may not be able to provide a direct response to all submissions. We thank you for any feedback you can provide.

E-mail: feedback@designingrisk.com

Reader Services

Visit the book website

http://designingrisk.com

or

http://itaseries.com

Register this book for convenient access to any updates, downloads, or errata that might be available for this book.

CHAPTER 1

Overview

"Everything in life has some risk, and what you have to learn, is how to navigate it."
--Reid Hoffman (Co-Founder of LinkedIn)

Risk is often an overlooked part of the design process, or an afterthought. This chapter shows how the process of understanding risk is interdependent with other design inputs. A high- level view of the methodologies used in risk analysis are provided.

Figure 1-1 A Yorkshire pig, contemplating risk at a smokehouse

Everyone has heard of Murphy's Law; anything that can go wrong will go wrong. This adage can be understood and applied in a variety of ways.

First, with a defeatist attitude, that the universe is conspiring against you and that there is nothing that can be done or undone. As in, "It's Murphy's Law that I got sick on my vacation".

Secondly, it can be taken in a negative light when referring to others, based on their skills, or work ethic, etc. As in "Sam set the coffee maker on fire again; Murphy's Law!" The third context can be seen as a call to action, to dig deeper.

What does "anything that can go wrong" encompass? What is the scope of this "law"? What is it that can go wrong? The limit of possibility is only bound by our imagination. Put on the hypochondriac's hat, and see where it leads.

Because I have two kids, I regularly have to break down the complex to its more palatable components. Often, I will take something that's familiar and twist it slightly to get the wheels turning in their heads. This also helps me simplify a concept that has many appendages to something that is more tangible. It can also make a heavy topic lighter with some levity.

In this light, I'll start with a tangent on the children's story of the three little pigs.

> The Three Little Pigs is a story about three pigs whom built three houses of different materials. A big bad wolf blows down the first two pigs' houses, made of straw and sticks respectively, but cannot destroy the third pig's house, made of bricks.

Here we have a scenario where three individuals approached a problem in different ways. "Pig 1" saved on cost and effort and made his house out of straw. The benefits are apparent; simple modular construction, cool in the summer and it keeps the rain off. There are a lot of people who build straw bale houses.

If "Pig 1" wanted to weigh out his building options, I would have suggested looking at the whole picture; environment, climate, regional census data, police data showing the average number of break-in from wolves per capita, etc.

If "Pig 1" did his homework, then he would also have a list of known wolves in the area, wolf early warning systems, decoy houses with wolf traps in them, and maybe even get to the root of the problem and start a wolf food bank or harm reduction outreach center. "Pig 1" could even draft a formal agreement between the wolf packs and the pig herds.

But instead, "Pig 1" was lazy and opted to design an environment that would fail easily, catastrophically and without failure in mind.

"Pig 1" is complacent, negligent, and lacks foresight. "Pig 2" and "Pig 3" aren't much better in this regard.

This book shows you how to not be a pig.

As an Architect or Administrator in IT, there are many ways of doing things that conceptually achieve the same result. However, the way in which something is done is often more important than the end result. This is a tradecraft, an attention to detail, form and function, and an appreciation for the subtleties of design that really make the difference between a ROLEX and a ROLLEX (the knock-off). Asking the question of why something is built in a certain way will often lead you to the answer of why it is not built another way.

Let's take a different approach to a topic that is considered a left-brain activity, centered on compliance and threats of hackers and service level violations. Let's assume that this book is a thinking book that takes you so far out of the box, that the box will be nothing but an old memory.

1.1 Risk is a good thing. Why we need to understand it.

Risk is often seen as a bad thing. It can be if it is ignored, or not understood properly. However, it can also be a tool to make good decisions. You can understand it by considering cause and effect, with long term and short-term outcomes.

In chaos theory, the "butterfly effect" dictates that small changes in initial conditions can have massive changes down the road. In theory, if a butterfly flaps its wings in one part of the world, it may have changed the path of a hurricane in another part of the world. This concept is used to illustrate the long-term ramifications of the smallest of activities.

A common colloquial adage is: "Hindsight is 20/20". Meaning that looking to the past, your vision is perfect, but to the future it is blurred.

Now if you combine the concepts, you can see that looking back small changes then could have made massive changes in the future. What if it were possible to do this? To understand the chain of cause and effect and the possibilities of outcomes. You could then make small changes to achieve those outcomes that are most desirable or avoid those that are detrimental.

A risk is the inverse of an opportunity. By understanding risks, you are creating opportunities. Let's break down what a risk is first:

Risk is the possibility that something detrimental may occur.

It has an impact on something and a probability that it will happen.

These two variables determine how much stock people put in that risk, and the effort used to respond to it. The combination of them is called severity, which can be seen in a risk matrix. We'll get to that in a minute, but first let's see where risks fit into the design process.

1.2 Constraints, Assumption, Risks, Requirements, Dependencies (CARRDs)

These five things are known as design inputs: constraints, assumptions, risks, requirements and dependencies.

It is from these five that we can the glean details of the design and formulate our decisions on how to realize them.

[side note: Some other combinations of these inputs form other acronyms that you may come across. Some of these are RAID (Risks, Assumptions, Issues, Dependencies) and ACRID (Assumptions, Constraints, Risks, Issues, Dependencies). These are used in project management and not as relevant for design, as they do not include requirements.]

Constraints (or also known as non-functional requirements) are the limitations of how we can do something. This may be something like having to use a particular software build, or being forced to use a limited bandwidth connection for site-to-site replication.

Assumptions are the missing pieces of the puzzle that have not been defined yet, but need to be defined in order to increase accuracy in planning and design. An example is this; Person-A will wear a red shirt tomorrow. Whether this is true or not, is not the point. Sometimes this is completely made up, but often it is gleaned from experience, trends, recommended practices, or from just a good understanding of how a specific scenario will unfold. If Tuesdays are red-shirt days historically or if you put a bucket of red dye in the laundry of Person-A, then the assumption is more likely to be true. Assumptions are not random, they are there to round out the design. Better examples would be; that all devices in the network will share common DNS servers, or that there are no rats in the ceiling that eat fiber-optic cables.

[side note: Rats chewing through cabling is the number one cause of service interruptions tied to technical problems for the cabling industry. Dow Chemical has identified 2000 species of animals that are fond of cables (FOC). Some companies have started embedding capsicum in the PVC sheathing of the cables to deter animals in a humane way. Larger companies like AT&T and Qwest use corrugated metal sheathing to protect their cabling.]

Risks are what this book is about. They define the cause and effect for doing something one way or the lack of doing something. For example, is the inherent risk in the reliability of a new and unproven technology. Using it in production before due diligence and regression testing validates it could cause widespread failure of all interrelated systems. The probability of this depends on the amount of effort and scope of testing that has been performed beforehand.

Requirements are the key points provided by the stakeholders that must be adhered to. If there is one thing that molds a design more than anything, it's the requirements. If all the requirements defined are not met in the design, then it is a failed design. However, if all the requirements cannot be met, or should not be met, then this needs to be called out right away and the design must be modified to address it. Sometimes this is not determined until certain avenues have already been traveled, and new information may come to light that changes the feasibility of the requirements. This possibility in itself is an underlying risk to just about every design.

Dependencies are the possibilities that are defined based on key happenings or interactions between events, objects and actors.

One way of using dependencies is to state that for an intended specific result to occur, specific interactions need to occur. This is the simplest method and often used in project management to achieve the end goal (this is called the waterfall method). For another example, you can think of package dependencies in Linux. With this method, if an unforeseen or unexpected interaction occurs that changes the project timeline, scope or budget, then the project must adapt and the plan must be modified.

Another way of using dependencies would be if during the building of a bridge one side settles or slips. This would force a complete structural re-evaluation. Where the risk is that this may occur affecting scope, timelines, safety, etc. The dependency is that if this does occur, then the re-evaluation will occur. This may then turn into a rebuild under tighter scrutiny, or it may be salvageable, or it may be deemed that the cost of salvage is greater than the cost of a redesign and rebuild. The potential realities that fork off from this occurrence all have ramifications. This is where those are defined.

With the latter method, it requires much more foresight, analysis and planning but it is inclusive of all foreseeable eventualities, their consequence and resulting actions.

When putting all of these together, the cohesion solidifies many variables and allows for a detailed vision of the design to emerge. Now it's time to write down all the key points learned that would enable you to build the best design.

1.3 Design Decisions

The more design decisions that are defined, the better. They will refine and sculpt a concept into a tangible form. For larger designs, it is not unheard of having 20+ pages of design decisions. By writing down the reasons for doing things a certain way, the justification can be validated. Sometimes people will be opposed to the way an environment is structured solely because it is different than what they expect, but don't know why. With the decisions defined, the thought process and reasoning is explained. This can be very helpful in winning over old dogs that have done things the same way for 20 years and are averse to change. This does not apply only to individuals, but also to conservative organizations that have a culture of keeping things the same unless substantial justification is provided. This also important for future people to understand why something was done before they arrived at the company. If you read the first book in the IT Architect Series (Foundation in the Art of Infrastructure Design), then you will be familiar with the requirement and constraint tables. This is the logical extension to those.

The key points of a design decision are these fields:

1> Problem Statement and Business Goals

2> Design Decision

3> Justification

4> Alternative Options

5> Implications

The simplest way to manage the documentation of Constraints, Assumption, Risks, Requirements and Design Decisions is to have a spreadsheet that contains them all. One area per worksheet is best. Then when the spreadsheet is complete, simply copy the table into your document.

For larger initiatives, this is not feasible, and a design register must be used to contain all aspects of the design process, including the previously stated CARRD inputs and a document management system with revisioning. Every CARRD input and design decision output will have an ID associated with it. The prefix of the ID will indicate the type of input or output.

An example would be:

C001	The prefix start ID for Constraints
A001	The prefix start ID for Assumptions
RI001	The prefix start ID for Risks
RE001	The prefix start ID for Requirements
D001	The prefix start ID for Dependencies
DD001	The prefix start ID for Design Decisions

An example Design Decision output would look like this:

ID	**DD001**
Problem Statement	In a VMware vSphere environment what is the most suitable type of NIC teaming to use in order to ensure maximum utilization of the teamed interfaces without a dependency on specific physical switch configuration?
Design Decision	Load Based Teaming for port groups on non-aggregated links when using a single, or multiple switches on a single Ethernet fabric.
Justification	Minimal configuration is required. It makes use of links based on load, leading to overall higher throughput. No physical switch configuration is required. It provides the best balance of performance and availability compared to the alternative options.
Alternative Options	Route based on originating port ID. Route based on IP hash. Route based on source MAC hash.
Impact	Overall bandwidth utilization capability is increased. Enterprise+ license requirement as VDS is required.
Risks	Financial constraints may limit license level in environment. Hosts with fewer uplinks may not be able to associate the same number of uplinks and thus teaming capabilities will not be identical across all hosts.

1.4 Conceptual, Logical and Physical Designs

For those that do not work with the design process on a daily basis, there is a tendency to try to simplify the separate conceptual, logical and physical models into one. Sales people love doing this because it makes their life easy. They simply have a product and then hunt for a fitting problem for it to solve.

The product SKU or "Stock Keeping Unit" is the internal ID of a product that has been completely defined by a vendor or manufacturer. It simplifies the delivery process of solving problems. An example would be if someone wanted to buy a light fixture for their house.

The "Owner": **CONCEPTUALLY** thinks, "I would like lights on my patio about 30 feet into the yard. They would be for illuminating the yard at night, but I want the lights to be bright and have no messy cabling that would detract from the ambiance."

The "Architect" thinks: "The situation allows for two **LOGICAL** alternatives.

Either:

You would need to have a pedestal lamp about 8' tall that would match the style of the patio. It would be solar powered with the cell and charge controller built into the lamp itself.

Or:

You're second alternative is to have lights connected to the house wiring, but hidden underground so as to not detract from the ambience of the patio.

The architect recommends to go with the solar option because it minimizes in the installation costs since it's so far from the house.

The "Builder" counters: "The Architect is suggesting solar lamps that would be 8 feet high and provide enough light to illuminate the patio. The problem is that the patio is in a windy area, so we would need to anchor it to the ground. We could **PHYSICALLY** build an anchored base to place the standing lamp in. Your other alternative is to have it custom made directly from a discount manufacturer and ship it with a self-anchoring base."

After the lighting solution was installed, the designer now has all the parts, models and process for the build documented. This entire solution can now have a SKU assigned to it by the company to make for ease of reselling.

Here is where the problem lies: if the requirements are 100% identical, then the solution can be sold with this **PHYSICAL** design. However, if parts cannot be sourced, or there are circumstances that do not allow for the exact same build, then the **LOGICAL** design will need to be looked at to come up with a new physical design.

A conceptual design is the high-level overview depicting the requirements.

A logical design will not change very often and can be used to convey a detailed solution for a conceptual design regardless of parts availabilities, market changes, model changes, versioning, etc.

A physical design is very detailed and is only usable if there is no change in any aspect of the CARRD inputs. It is very repeatable and only requires the deployment team to follow a detailed build guide.

1.5 Risk Governance vs. Risk Management

How an organization deals with risk depends on a number of factors. These could include organizational obligations to third parties, industry practices that must be adhered to, or security strategy changes after an incident occurs. Management consumes the factors, and then creates a mandate that needs to be followed and a timeline associated with it.

The mandate is then delivered by a set of controls that defines it. These controls are essentially the rules by which an organization must adhere to be compliant.

This process is called Risk Governance. It is a top-down approach that defines, frames, analyzes and perpetually reviews the environment for compliance against defined controls.

An organization requires a team of people for this. It is not a small endeavor to perform properly, nor is it quick to implement.

Risk Management is the process of determining what may impair the realization of an organization's business objectives, the impact and the probability of it occurring. The analysis of the risks will then have actions associated with them to prevent or reduce their impact or likelihood. Whether these actions are performed depends upon other factors such as cost and time. An organization will have to weigh risk severity, effort required in resources and cost associated with the action. It would be wonderful if all organizations could afford to address the risks they identify, but that is not the case.

Often it is only a matter of plugging major holes or putting out fires. The smaller, but less critical actions will get put on the shelf until they change their criticality, effort or cost to fit in line with what the organization is willing to do. This is called risk acceptance.

Risk Governance as a practice is always inclusive of Risk Management, but the other way around is not true. Risk Management can exist solely within a security framework, or under the purvey of a business analyst, depending on the type of risks that are included. Risk Management is only a smaller subset of the greater picture, but can have great benefit when implemented. An example of this is a Security Operations Center (SOC) that may care about vulnerabilities and threats to digital resources, but it is not under their scope to care about the risks associated with market trends, implications of sales projections or threats to physical resources such as power, facilities or personnel.

In this way, without a dedicated Risk Governance practice, some of the roles and oversight it provides is distributed amongst a variety of other departments or divisions. The side effect is a possible lack of cohesion, overlapping efforts, or gaps in overall effectiveness in risk related strategy.

Both Risk Governance and Risk Management have defined guidelines that must be adhered to in order to be effective. Whether or not they are being adhered to is what is referred to as Compliance. Often, the combination of these three components is referred to as GRC (Governance, Risk Management and Compliance).

[side note: This can get confusing when there is Governance of Risk Management, which is the practice of ensuring that only the most important Risk Management actions are taken. This is different than Risk Governance, which creates the overarching policy and framework.]

1.6 Risk Appetite

Risk Appetite is the concept of how much risk an organization is willing to endure before the potential losses outweigh the anticipated gains. When developing a Risk Governance practice, the definition of the Risk Appetite is important. It will factor into every decision that the organization will make, as an undercurrent. It reflects on the perception of management from stakeholders and the market, it shows some business policies and practices, and it influences the culture within an organization.

A high-risk culture could mean high capital gains, insightful leadership, and plentiful revenue that can tolerate the possibility of misstep or failure.

It could also mean that the company is in its death throws and needs a big winner to stay afloat. Or it could be reflective of mismanagement and a warning sign of potentially grave losses.

Regardless of the truth, the external perception will be colored by how successful the organization is in the end. A poorly run organization that got lucky with a gamble may be seen as a leader and groundbreaker, even though they could have drastically failed.

On the other side of the spectrum, a well-run organization that is conservative in its risk appetite may be seen as inflexible, and thus lose market share to the less risk-averse movers and shakers.

The definition of a risk appetite can be done in four steps.

1. Review the organizational objectives.
2. Identify the current risks to those objectives and what can be endured before the operational strategy must be changed.
3. Define the thresholds that would force strategy change if they were to occur.

Formalize the findings into a document that becomes part of the overall organizational strategy.

1.7 Risk Awareness

Risk awareness is the practice that an organization will use to ensure that everyone is participating in the initiatives to protect the business objectives. If a risk awareness plan is not in play, that leaves the organization open to things like social engineering attacks, the ramifications of poor planning, cost over-runs and technical screw-ups without mitigation plans. The list can go on and on.

Without a risk awareness strategy in place, all the work that is put into risk governance and management could be undone in a minute.

A fortress with a broken back door is weaker than a shed with a padlock. If the people within an organization do not follow policy, the entire organization is at risk.

Risk awareness is the process that will show personnel why they should follow policy, what the policies are and how to be vigilant in the face of breaches of policy.

A poor risk awareness program could backfire on the governance goals. If staff are choked and drowned in red tape, bureaucracy and rules, then they will simply find a work-around. This is especially the case when their day-to-day jobs require that they do something that they are not allowed to do.

Here is an example:

An employee (let's call him Harold) is required to take photos for insurance investigatory purposes. He has done this for many years on film, then digitally. The process has always been to convert them to a suitable standard format (TIFF, PNG, JPG) then attach it to a case file, add metadata (notes) and then associate it with the file library.

Harold has always taken the pictures on his own camera or phone, scanned and emailed them to himself, then from the email downloaded the attachments at work, then uploaded them to the office share folder and add a notes file in the same folder.

Then a security mandate came down from management with these controls:

a) No access to external mail systems from the office.
b) No remote file transfers.

No USB drives attached to office PCs.

No outside devices allowed on the internal network.

The new method required Harold to do the following:

    a)   Take the pictures with a work provided camera

Upload the pictures to a secure portal.

Add notes on the secure portal.

To management, this sounds like a security win and a streamlined process. The practice for the Harold was not as smooth.

- The provided camera was prone to crashing and did not work as well as his personal camera on his phone.
- Access to the secure database required two-factor authentication over a VPN and from a small set of remote IPs.
- The VPN had performance issues that were never addressed by internal IT.
- The secure portal required a specific version of Java to run the client, and it was also intermittently unreliable.

Because of these factors, Harold was now taking 4 times as long to do the same job he was previously. He was getting complaints from management and missing deadlines. He decided to resolve this by doing the following:

- Use his camera phone to take the sensitive pictures.
- Install a RAT (Remote Access Tool) on his computer in the office.
- Install an app on his phone to access the computer at the office.

Upload the pictures directly to his computer in the office.

Use the secure portal client from the computer in the office (remotely).

Now Harold is patting himself on the back and management is saying "good work" on catching up with the workload.

What happens next, could have been avoided. Harold forgets his phone at a coffee shop. Someone finds his phone (let's call them Jay) and says "Hey great! Let's see what's on it". They find the RAT app and connect to Harold's computer remotely, which he left unlocked to speed up login times. They then show a friend (let's call him Samuel). Samuel thinks this is k-rad hacker stuff and installs the same RAT app on his home machine so he can hop onto Harold's computer any time without the phone.

Jay wipes the phone and sells it on Craigslist for $50. This gets him some Chinese food and couple of coffees.

Samuel, being the not-too-security-conscious person that he is, happens to have some spyware and a key logger on his computer. That key logger is sending everything he types, the applications he is using, screenshots, timestamps, etc., to a hacking group called Legion of Lulz (LoLz).

LoLz then logged into Harold's computer, put in their own RAT, removed the original RAT, changed passwords, accessed the office shares and encrypted the entire file system of the file server in the office. They then sent a ransom note demanding payment of 1 bitcoin from every contact on Harold's contact list (on his secure office email), threatening they would delete all the data if demands were not met.

Now Harold is not patting himself on the back, and neither is management. All productivity is completely halted, the office is shut down, Harold is fired and under investigation, the "hack" is on the front-page news of the papers, and the executive board pulls the Chief Security Officer in to ask, "how could this happen with your new Risk Governance strategy in place?"

Risk awareness is not just about top down directives, but rather should be a shared responsibility for both employees and management to keep the organization safe. In this scenario, we could put some blame on just about everyone. Blame does not solve anything though, prevention and education does.

1.8 Assessment and Identification

When it is determined that a risk-centric initiative is taking place, one of the first and main functions is to perform a discovery and identify all risks from every angle.

An assessment often begins with a conversation with stakeholders. This then turns into defining the standards that will be followed, the language that will be used and the controls that are used to map actions to. Sometimes organizations will have to use multiple standards, which have an overlap. In this case, the mapping of controls from standard to standard needs to be documented. In the event of an overlap, the standard that has a control that is more inclusive will be the one that is used.

After the controls are defined, the means of assessing compliance needs to be determined. Once they are defined and assessed, then a compliance gap document is created along with a remediation action plan.

This process will now be the ongoing compliance management that becomes part of operations. The schedules for reviewing control compliance are defined and reviewed by the Risk Management team and vetted by management.

In some cases, the defined controls are enough for the auditors. Other situations demand a more comprehensive and proactive approach. This would involve a full risk discovery process that can put together all the possible things that can go wrong into scenarios that prove their probability and impact.

1.9 Risk Scenarios

These are very descriptive "what-ifs" that go into as much detail as possible. They will map out dependencies, dependent processes, and probabilities. Scenarios will involve highly structured flow charts that show what will happen based on things such as known and unknown variables and risk factors.

The components of a scenario are:

1. Time
2. Event
3. Threat Type

Resource / Asset

Actor

Analyzing scenarios allow for a broader view of the infrastructure over time. Things that may have been hidden, or unforeseen become uncovered. Events that are inevitable became blatantly apparent through several scenario flows.

Once a risk scenario is deemed to be relevant and realistic, then it will be added to the risk register for the organization. This is a spreadsheet or database that contains a record of all the risks postulated in an organization.

1.10 Risk Factors

A Risk Factor will increase or decrease the frequency or impact of a Risk Scenario. This would include organizational capabilities and, or internal and external environmental factors. Some example of these would include:

a) Market pressures
b) Geo-political situations
c) Strategic Priorities
d) Risk Management Philosophy / Risk Appetite
e) IT capabilities (resources, knowledge, effectiveness)

Risk Factors must be taken into account when creating Risk Scenarios as they can change the classification of a Risk from a low level to a high level. Thus, ignoring a Risk Factor or not acknowledging it is a risk in itself.

1.11 Risk Impact and Probability

If we create a scale of 1-10 of how likely something is to occur and then another scale of the impact from 1-10, then we can make a grid to show where those risks reside.

If we were to get very granular, then in theory, we could make a 10 x 10 grid and have 100 separate zones. The problem with that is it becomes difficult to manage 100 potential actions for a single given risk.

Another method would be to create quadrants and then only have 4 possible actions for a given risk. Though this method would be better than none, it relies on the quantitative analysis of the risk to be very exact.

This process of creating zones and assigning values to it is called the Risk Matrix.

In this book, we will use a 6-zone Risk Matrix model, which accounts for transition areas and makes allowances for less than perfect quantitative analysis, while having the fewest zones possible.

The zones are organized alphabetically (A, B, C1, C2, D, E) from lowest risk to highest.

Each zone encompasses a spectrum of 2.5 units on the scale from 1-10.

The exception to this is C1 and C2, which are considered to be transition zones, as they are so close to the center axis of the grid. C2 in particular is very close to the 5/5 point, which could change the actions and resources associated with it dramatically.

In this zoning, I = impact, and P = probability.

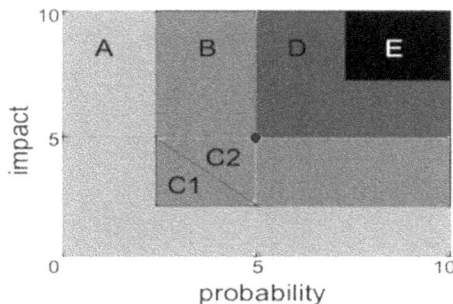

Figure 1-3. A six-zone matrix model

A is considered low risk, based on the combination of both impact and probability. This zone encompasses I-0 to I-2.4 vs P-0 to P-10 and I-0 to I-10 vs P-0 to P-2.4

B is considered moderate risk. This zone encompasses I-2.5 to I-4.9 vs P-5 to P-10 and I-5 to I-10 vs P-2.5 to P-4.9

C1 is considered a low moderate risk transition zone. It encompasses I-2.5 vs P-2.5 to I-4.9 vs P-2.5 to I-2.5 vs P-4.9

C2 is considered a high moderate risk transition zone. It encompasses I-4.9 vs P-2.5 to I-4.9 vs P-4.9 to I-2.5 vs P-4.9

D is considered high risk. It encompasses I-5 to I-10 vs P-5 to P-7.4 and P-5 to P-10 vs I-7 to I-7.4

E is considered extremely high risk. It encompasses I-7.5 to I-10 vs P-7.5 to P-10

By classifying the risk, you can determine how to respond to it and what sort of resources you allocate to do so.

The amounts of resources that are available to address risk are not limitless, and usually have to be justified to management and the executive team on a regular basis. An example of this is the resources allocated to a DR site, even though there may have not been a site failure for many years or at all.

By using the classification method, the justification is quantified, verifiable and difficult to refute.

1.12 Scorecards and Key Risk Indicators (KRIs)

Key Risk Indicators (KRIs) are similar to Key Performance Indicators (KPIs) that are used by management to determine a quantitative value of how an organization is performing against its objectives. The difference is that KPIs perform analysis on data that has already occurred, whereas KRIs evaluate situational changes on many fronts to determine the future likelihood or emergence of a risk.

Often these KRIs can be external indicators that coincide with internal events.

An example of this would be how the fluctuation in pricing of rare metals affects the technical capabilities of a nation in a specific area. If we look at the metal Germanium, we see that historically it was used strictly as a semiconductor in the production of solid-state electronics. However, because of short supply globally, the production process of creating silicon was improved and became more cost effective than mining or recycling Germanium.

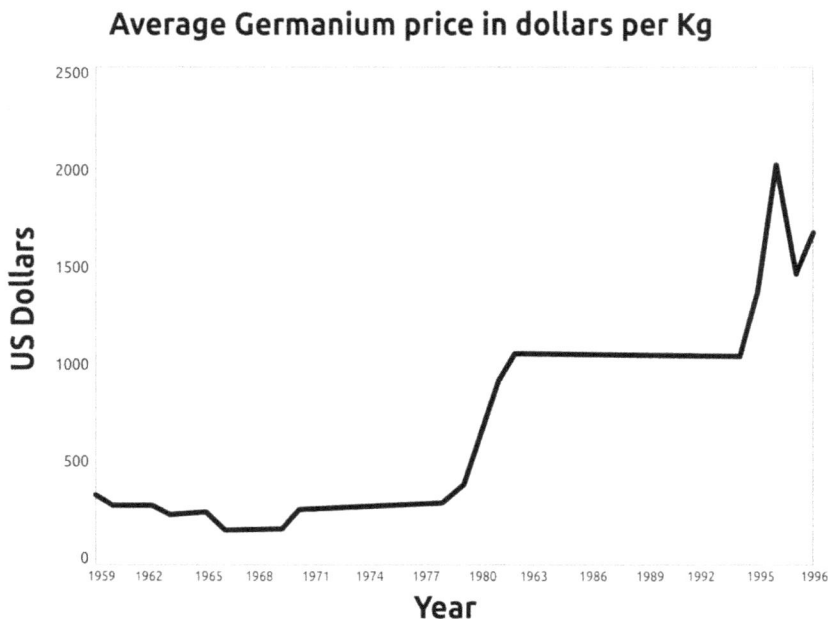

Figure 1-4. The indication of two massive jumps in the price of Germanium in recent history

(reference: http://minerals.usgs.gov/minerals/pubs/metal_prices/metal_prices1998.pdf)

[side note - Significant events affecting germanium prices since 1958
1979-82 Increased demand, tight supply
1984 National Defense Stockpile (NDS) authorization, goal 30,000 kilograms
1987 New authorized NDS goal of 146,000 kilograms
1991 NDS goal lowered to 68,000 kilograms
1996 Increased demand, production shortages
1997 NDS stockpile sales begin]

Other uses were found for Germanium, such as a doping material for fiber optics and the production of high-performance solar cells. When these discoveries occurred, the demand increased and thus the price, because of the shortage. Two time periods in recent history saw the price of Germanium double within a two-year period and never fully drop back down. First this occurred in 1980-1982, then again in 1995-1996.

If a nation wanted to stay "state of the art" in the fiber-optics industry during this time, they would have to have deep pockets to subsidize companies within the nation to obtain the Germanium from other countries for the raw materials, or they could stockpile it before the large demand was there.

The latter is what happened. Based on the technological advances of the time and the raw materials to capitalize on them, the future speculative price for Germanium was a KRI directly related to the effectiveness of a nation's ability to capitalize on new technologies.

So around 1980 and 1995, the world started creating stockpiles of raw Germanium ingots and prices soared.

1.13 Risk Response

Once a risk has been defined, validated and entered into the risk registry, how it is dealt with goes back to the organizations risk appetite. The Risk Matrix shows the severity based on the six zones that encompass probability vs impact. Based on the severity, the organization will respond to the risk in one of four ways:

1. Avoidance
2. Transfer
3. Acceptance
4. Mitigation

I will use an example to exhibit each of the four responses to the same situation.

A smartphone manufacturer found that batteries installed in a certain model might explode and cause a fire if overcharged or damaged. The tests show that this may happen in 1 out of every 100,000 phones. The fire could cause first degree burns and possibly death from fire, or by distracting someone while driving and cause a fatal crash.

Avoidance removes the risk from the situation. It does not solve it, address it, or mitigate it. An example of this kind of response to the scenario is:

The smartphone manufacturer recalls all phones of this model produced and removes it from future production.

Transfer moves the risk to a third party. It is often an option that has caveats or additional costs. An example of this kind of response to the scenario is:

Installation of the batteries is subcontracted to a third party, which takes full liability for the battery failure. Another option under this form of response would be to have manufacturers leverage their product liability insurance. This is not something that can be done in IT within an organization. However, it can be done with projects with clients through collaborative outsourcing if it is explicitly defined in the contract.

Acceptance is used when the impact is low or when no other form of response is usable. There are two kinds of acceptance, active and passive.

In this example, the **Active Acceptance** response to the scenario is:

The smartphone manufacturer recalls all phones of this model produced, repairs the battery and replaces it with another validated one for future production.

In this example, the **Passive Acceptance** response to the scenario is:

The smartphone manufacturer awaits an incident, contacts the affected parties and settles quietly out of court and the public eye.

Mitigation is used when the organization chooses to acknowledge the risk and determined that the severity was past the threshold of acceptance. Mitigation is proactively used to lessen, or eliminate the impact of the adverse effects of the risk.

It can be considered a checklist of tasks to perform during every stage of a project, or situation.

In this example, the responses could be varied, depending on the mitigation capital. As the name implies, it is used to prevent risks from occurring by having the foresight to estimate possibilities.

- They use top of the line components without trading of cost for safety.
- Best practices for design and build are followed as per industry standards.
- They perform standardized failure analysis / QA of all systems.
- They have a comprehensive action plan for recalls in the case that something was missed during testing.
- They have a law firm on retainer for product malfunction lawsuits.
- They have a law firm on retainer for out of court settlements.
- They have an outreach group to offer compassion to the families of the affected.

They have a PR team dedicated to marketing spin and media relations.

A key thing to note is that the undercurrent to all risk response strategies is cost. Does the response justify the cost? Does the executive team endorse the response initiatives based on the organizational risk appetite?

The risk response also has an element of morality tied to it. It may be less costly to let thousands of people burn to death from exploding batteries and drag on the legal proceedings so that the complainants cannot afford to continue. Or it may be simpler to payout the families on a case-by-case basis with quick low-ball settlements. These may be the cheapest options for the company, but they do not paint them well in the public eye.

This scenario is similar to the dialogue in a scene from the movie **Fight Club:**

Narrator: A new car built by my company leaves somewhere traveling at 60 mph. The rear differential locks up. The car crashes and burns with everyone trapped inside. Now, should we initiate a recall? Take the number of vehicles in the field, A, multiply by the probable rate of failure, B, multiply by the average out-of-court settlement, C. A times B times C equals X. If X is less than the cost of a recall, we don't do one.

Business woman on plane: Are there a lot of these kinds of accidents?

Narrator: You wouldn't believe.

Business woman on plane: Which car company do you work for?

Narrator: A major one.

1.14 Enterprise Risk Management

Enterprise Risk Management, or ERM is a discipline that defines risk within enterprises from all sources such as:

- Operations
- Strategy
- Finance
- Physical hazards
- Politics

Some of the goals of ERM are the following:

1. Establish a common language terminology that can be used for internal discourse and education.
2. Identify current organizational risk state in each key area.
3. Establish a risk committee.
4. Create the risk register.
5. Create the risk matrix.
6. Define the organizational risk appetite.
7. Associate ownership to each risk.
8. Define KRIs for the organization.
9. Develop response plans for each risk in the register.
10. Create an auditing role within the organization to ensure compliance.

Create the Risk Governance framework for the organization.

The oversight and ownership of ERM usually resides within the executive branch of an organization. Often it is instantiated by the CFO (Chief Financial Officer) or CSO (Chief Security Officer). Some organizations such as insurance companies, have a dedicated Chief Risk Office to oversee ERM.

1.15 Chapter Summary

1. Constraints, Assumptions, Risks, Requirements and Dependencies are inputs to a design.
2. Design decisions are an output, and require justification.
3. Conceptual designs are high level and used to convey an idea.
4. Logical designs have no products or parts explicitly referenced and are used to show how something will work. They do not expire.
5. Physical designs reference specific parts, vendors and versions. They are used in the build / deploy phase and they do expire as new models, firmware, version, etc. are released.
6. Risk Governance is a top down approach that is all inclusive.
7. Risk Management has guidelines that must be adhered to, and is often practice (or department) specific.
8. Compliance ensures that the guidelines proposed are being followed.
9. Risk appetite is how much risk an organization is willing to accept before the potential losses are unacceptable.
10. Risk awareness is how the Risk Management policies, procedures and reasoning are socialized.
11. Risk assessment is the process that's used to determine which areas of an organization that are at risk.
12. Risk scenarios are the details (who, what, where, how, etc.) of how an incident may occur.
13. Risk factors are things that increase or decrease the impact or frequency of a risk scenario.
14. Risk impact and probability are used to determine criticality, which is mapped on a matrix.
15. Key Risk Indicators (KRIs) are external indicators that give insight into changes in criticality.
16. Risk Response is what is done when an incident occurs.
17. Enterprise Risk Management is the discipline that encompasses all aspects of risk in an organization.

1.16 Chapter Review Questions

1. What is the difference between a functional and a non-functional requirement?
2. Give an example of a KRI in relation to electrical power?
3. True or False? A high-risk appetite is often required for rapid growth in a short period.
4. Which Risk Response requires the least amount of effort for the desired result?
5. Which Risk Response is the most expensive? Why would it be used?
6. How does Risk Awareness affect the success of a Risk Governance strategy?
7. Within your IT infrastructure, what risks influence your design choices for risk mitigation?
8. What are the levels of risk that require investment to address?
9. What Risk Response are applied for the risks identified within your infrastructure? Remember to include internal and external risks.

CHAPTER 2

Indestructible

"Your mind is working at its best when you're being paranoid. You explore every avenue and possibility of your situation at high speed with total clarity."
--Banksy (Street Artist)

Is it possible to keep an environment running under any circumstance? What are the considerations? How many failures can you tolerate? How do you measure it? This chapter discusses these concepts and more.

2.0 What is indestructible?

This question can lead to philosophical discussions such as, "What is really physical if every bit of matter is simply a different arrangement of energy fields?" Energy is neither created nor destroyed, but simply changes form. With that line of thinking, if everything is energy, then everything is indestructible.

However, in this book what we are referring to less abstract and theoretical concepts. Specifically, we are referring to the change of state of a system from "operable" to "inoperable". When a system breaks and it is no longer able to function in its primary role and it cannot be repaired, then it is considered permanently out-of-service, or destroyed.

The truth is that a system doesn't break like a sheet of ice, or like Humpty Dumpty. A system is comprised of sub-systems, components, buses, electrical traces, software, processes, facilities, and people. The function of a system is how it's used, it's purpose. This is arbitrary and subjective. Two identical systems can have completely different primary roles, whilst two radically different systems can have the same primary role. This is possible because of standardization and commoditization of hardware.

Since standardized hardware is used for modern enterprise computing, the conversation becomes less about the hardware and more about what it enables us to accomplish, or in other words; the applications, services and processes we run. Every process requires an operating system (OS) to run on. The OS and the applications require a certain amount of resources from the CPU, memory, network and storage. The number of concurrent instances of an OS on a host depend on the efficiencies of the hypervisor and the resources available on the host.

When a system breaks and it is hardware related, it will be within one of those four areas (compute, memory, network, storage). Since there is that standardization of hardware in the industry, broken things can be replaced and everything can be repaired.

[There was a time in the not too distant past when software was coded for specific chip makes and models. A software developer had to have an intimate knowledge of both hardware and software. The code would be extremely efficient, made in machine language, and it would run lightning fast.

The main problem with this is that the code was very difficult to maintain when hardware changed. CPUs became a lot more complex and were capable of running many instructions simultaneously. Computer manufacturers then came to a crossroads; have a very efficient chip that can do a small set of things very well (Reduced Instruction Set Computer - RISC processor), or have a general-purpose chip that can do many different things at once (Complex Instruction Set Computer - CISC processor). RISC processors are more advanced, more efficient and use less power than a CISC processor. Apple Macintosh and PowerBook computers were originally using used CISC processors with the Motorola 680x0 CPUs. They then changed to RISC based processors when they moved to the PowerPC CPUs for the Power Macs., In 2006, Apple switched all Mac computers back to but changed to CISC based processors when they standardized on Intel processors. in 2006.

RISC processors have made a massive resurgence in recent years as we moved into the age of mobile computing. Applications have become king and battery life is a huge factor. That is why all smartphones, tablets, watches, etc., use RISC processors. You can effectively have the same applications, but with less power and using less CPU clock cycles.

But this book is about Risk, not RISC.]

If everything is repairable, then how can a system become destroyed? This is where we veer back to reality instead of vacationing in the hypothetical. True, every system can be repaired, but there is the question of what it will take?

If a marathon runner breaks their leg after a few kilometers into the race, can they finish the marathon? Broken bones heal, but not that fast. Time is a major factor involved with the reparability discussion.

If a baseball player gets hit in the head with a ball and is knocked out, does the whole game stop permanently? No, because there are extra players available to take their spot. This is an issue of resources.

If a space faring rocket is launched and crashes before the mission is complete, is the whole space program cancelled? No, because there is more money available to build a new rocket. This is a funding issue.

With a computing system, it can be considered permanently out of service if it cannot be repaired within a defined time period, or if the resources are not available, or if there is no budget to fix it.

When this happens, it can have far reaching effects like a company going out of business, or people dying in a hospital. So, if we wanted to make sure that a system never ever goes down, then the easiest way to do so is to remove the constraint of money. If money was no object, then it comes down to people, process and technology. How can a system be designed to ensure that the risk of going offline is effectively removed?

This chapter discusses all of the factors that need to be considered when constructing a highly available critical system with no budget constraints. In real life this never exists, because there are always constraints and trade-offs. There are always risks that can take a system offline, cause data loss, lost revenue, or have cascading effects on other systems because of interdependencies.

You can never truly remove all risk, but you can understand the effects of every design decision. You can accept quantified risk because you have accounted for it with a risk response. The architecture will reflect the organization's Risk Appetite, its executive leadership, and its people.

This is the basis of this book and it is what is meant by **Designing Risk**.

2.1 Resilient computing

2.1.1 Component Failure Rates

The concept of reliability theory was created in the 19th century for life insurance companies to determine the minimum amount to charge and turn a profit.

It has in turn been adapted to many other things besides the rate of mortality vs speculative profits. One such adaptation is the failure rate of components within a system in relation to the overall functionality.

You can break it down to a simplistic form by thinking of it like this:

An insurance company makes money when you're alive and paying for insurance. Based on all factors in one's life (health, risk and hazard in occupation and recreation, hereditary traits, etc.) the amount of money (premium) they can make while you're alive is calculated.

In systems design for a corporation, they make money while the systems are functioning.

Based on the reliability of components and subsystems, the amount of money they can make from the fully operational system, is calculated.

The same core theorems and math is used, but with different use cases.

Reliability is defined by something that is repeatedly measured and provides consistent results. This is slightly different than something that is considered "dependable". A **reliable** component is continually validated, whereas a **dependable** component has been validated at some point and there has not been enough data provided to counter that validation.

The latter is therefore somewhat subjective, but unfortunately it is the yardstick that is used by many when designing systems. This leads to a less efficient design, which will not necessarily be disastrous, but it will increase costs and resources in other areas such as repair cost and impact from unplanned downtime.

The test data of component failure has a few aspects that I want to call attention to in order to provide a better understanding of the process.

Censoring: To have 100% accuracy in understanding failure rates of a component, all components need to fail in the test. If there are survivors at the end of the observation period, then this skews overall results. Sometimes with components like memory, where they can operate at high loads for thousands of hours with failures in the parts per million, it becomes difficult to gather conclusive data. The two key aspects of a successful failure rate test are:

1. Being able to assess the failure points of the entire sample.

Determining the exact cause of the failure.

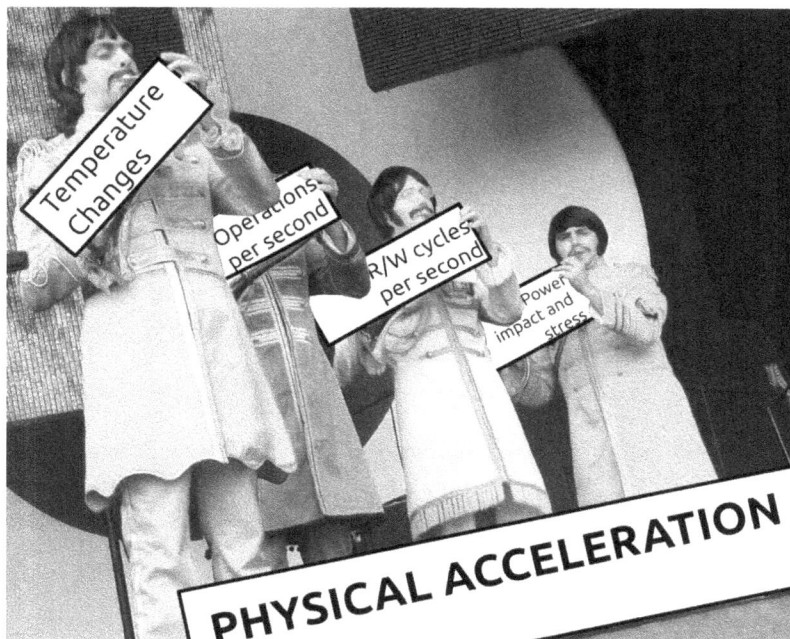

Figure 2-1. Physical acceleration gets by with a little help from its friends. Sgt. Peppers failure analysis club band.

The first aspect is difficult under normal circumstances, so sometimes a little help is needed.

When a component is operated at stress levels that are higher than normal, it is called physical acceleration.

This will decrease the time it takes for a component to fail.

It would be similar to if somebody went binge drinking, or if they went for a leisurely walk at a park in Chernobyl.

With components operating at a higher stress (i.e.; voltage or heat increase, higher duty cycle, more read writes per-second, etc.) the effects of a longer test period can be simulated until all components fail and the rate of failure can be accurately documented.

If an organization does not have the means to conduct empirical testing of their components, then they must rely on third party databases. A side effect of this is that unless the third party is financially motivated, the data will not always be current. If it is not a 100% reflection of components used within an organization, then the data will not be as relevant in calculations.

Some third-party databases of note are the "Computer Failure Data Repository (CFDR)" maintained by USENIX (https://www.usenix.org/cfdr) and the Hard Drive reliability report by Backblaze (https://www.backblaze.com/blog/) which is released quarterly.

[side note] - Changing the stress level can be seen as changing the time scale of the test period. The time-to-fail is multiplied by a constant (which would be the acceleration factor (AF))

We use the following notation:

ts = time-to-fail at stress
tu = corresponding time-to-fail at use
$hs(t)$ = failure rate at stress
$hu(t)$ = failure rate at use

Then with an acceleration factor (AF) applied, the relationships between stressed and normal use have the following relationships:

Time-to-Fail	$tu = AF \times ts$
Failure Probability	$Fu(t) = Fs(t/AF)$
Reliability	$Ru(t) = Rs(t/AF)$
Failure Rate	$hu(t) = (1/AF)hs(t/AF)$

AFs can only be used for one type of failure. For example, increasing voltage may increase failures due to heat or subcomponent failure, but it will not increase failures related due to things like excessive reads / writes / or erases.

So theTherefore, it is important that the use case and environment for the component should be defined before the AF is chosen.

For more math related to failures and risk, see chapter 6.

2.1.2 Manufacturing Processes and Standards

Historically, the quality of a product was determined by the skill of the craftsman who created it. It was a simple cycle that involved the design, sourcing of materials, creating the product, and ensuring it was made well. A skilled craftsman would do their own quality assurance, note any issues and fix them before it went to market. Each craftsman would also stamp his initials or seal into the product, proving credibility and providing accountability.

As the industrial revolution came in, mass production became more common, displacing the lone craftsman. To reduce costs, standardization of every aspect of work was created. In some cases, (as with the Ford Motor company) not only was work standardized, but so was family life and community.

By deconstructing work that would normally be done by skilled workers to its individual duties, unskilled laborers could be used to do a very specific task quickly. It was easy to train new workers and to keep a pool of laborers. Several mono-skilled laborers could perform the task of a single skilled craftsman, but more quickly, repeatedly and endlessly (with shift work).

[side note - This also had the side effect of intellectual property retention. No single worker was able to go and try to replicate the entire assembly line, because they had no knowledge beyond the single task they did.]

As manufacturing evolved, standardized methods of just about every task and process have been created. Every industry now has a standards group, or several applicable ones. The manufacturing chain has also become decentralized. No longer is it simply raw materials in, product out. Each product is made from many components from many manufacturers, sourced globally.

Look at the iPhone 6plus for example.

Figure 2-2. An exploded view of the iPhone 6plus.

From the image above there are a number of chips present, these are:

- Apple A8 APL1011 SoC + Elpida 1 GB LPDDR3 RAM
- Qualcomm MDM9625M LTE Modem
- Skyworks 77802-23 Low Band LTE PAD
- Avago A8020 High Band PAD
- Avago A8010 Ultra High Band PA + FBARs
- TriQuint TQF6410 3G EDGE Power Amplifier Module
- InvenSense MP67B 6-axis Gyroscope and Accelerometer Combo
- Qualcomm QFE1000 Envelope Tracking IC
- RF Micro Devices RF5159 Antenna Switch Module
- Skyworks 77356-8 Mid Band PAD
- Bosch Sensortec BMP280
- SK Hynix H2JTDG8UD1BMS 128 Gb (16 GB) NAND Flash
- Murata 339S0228 Wi-Fi Module
- Apple/Dialog 338S1251-AZ Power Management IC
- Broadcom BCM5976 Touchscreen Controller
- NXP LPC18B1UK ARM Cortex-M3 Microcontroller

- NXP 65V10 NFC module + Secure Element
- Qualcomm WTR1625L RF Transceiver
- Qualcomm WFR1620 receive-only companion chip.
- Qualcomm PM8019 Power Management IC
- Texas Instruments 343S0694 Touch Transmitter
- AMS AS3923 NFC Booster IC
- Cirrus Logic 338S1201 Audio Codec
- Bosch Sensortec BMP280

In short, there are 16 different component vendors for the creation of the physical device. To ensure that the product works at a high level of confidence, strict controls and standards have to be put in place at each stage of the manufacturing chain. Apple has hundreds of locations that it sources components from in 18 countries (http://images.apple.com/supplier-responsibility/pdf/Apple_Supplier_List_2015.pdf).

To ensure that quality is being met, manufacturing standards organizations like ISO (International Standards Organization) and IPC (Institute for Interconnecting and Packaging Electronic Circuits – now just IPC) are used to level-set the process and ensure consistency globally. Auditors are sent around the world to ensure that these standards are being followed.

Quality control is also implemented at the integration level to ensure that the build quality is as good as the component quality. In the end, a high-quality product is created affording a high premium price for this confidence.

When an IT solution is being designed, it will be comprised of many sub-components, each with different manufacturing chains and sources. To ensure that the end solution has a high confidence in quality, the integrator is ultimately responsible. How often is it that system integrators follow the supply chain of each component in the solutions they provide for clients? How often do internal IT departments do the same? Usually never. Then how is the confidence for the quality of the solution assured? Usually it is simply relying on the product vendors for QA and inter-operability. It is a transfer of risk and responsibility to the vendors.

This can be a nightmare when managing many vendors. One solution is to go with a single vendor that is ultimately responsible for conducting the QA and quality control on the solutions. An example of this is the converged infrastructure vendor VCE, which created the vBlock. Many different technologies go into a vBlock, but the solution is classified as manufactured by VCE because of the testing and standards that they have in place for QA. However, there is a financial premium to be paid for this assurance.

Another method is to validate the components and supply chain personally. This would entail having a catalogue of solutions, each comprised of specific makes, models, revisions, firmware, etc. A series of tests would then be done on these exact builds to determine the failure rate for the solution.

This is extremely costly and in a lot of cases, wholly impossible.

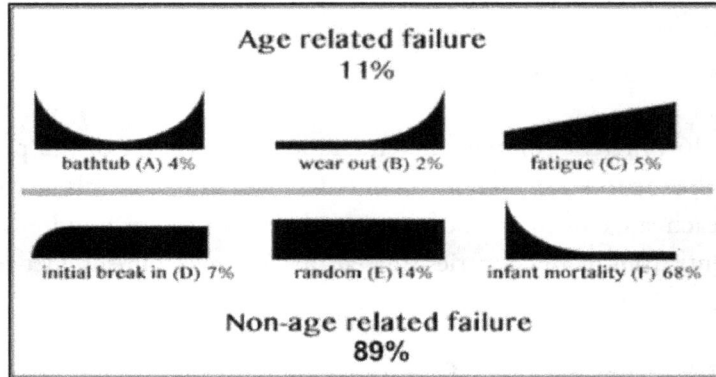

Figure 2-3. Failure patterns of components.

What can be done? If we look at nature, this is done instinctually by many animals. We call it "trial by tire" or "sink or swim" or "fly or fall". The weak will quickly perish.

It is stress tests and burn-in time.

What I am talking about can be seen more clearly in the following failure curves.

In these patterns from A to F, we see the time period in which components will fail on average.

It is important to note that these patterns and percentage values are not the same for every environment.

That being said, the infant mortality rate of components under stress is by far the greatest pattern of failure by percentage of total failures.

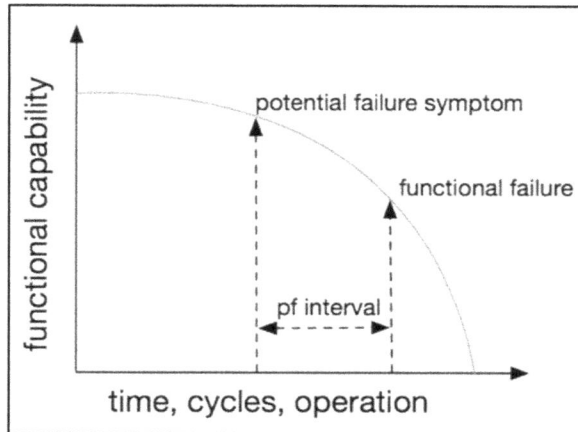

Figure 2-4. P-F Interval for component failure.

By performing a stress test, or burn-in test of a solution, you are more likely to find components that will fail early and replace them. The duration of the burn-in depends on the solution and the range in the infant mortality period.

The failure patterns above do not reflect total component failure, but rather negative deviations from normal operations. However, over time they will fail completely.

This period from initial symptom to total failure is called the P-F interval. The P-F stands for "Potential failure" to "Functional failure". It can also be thought of as the symptom to death period.

2.1.3 Environmental factors and failures over time

Age related factors on average only account for about 11% of Potential failure events. However, if the component failures have patterns similar to A or B (bathtub or wear-out), then you will have a large number of failures happening in a short period of time.

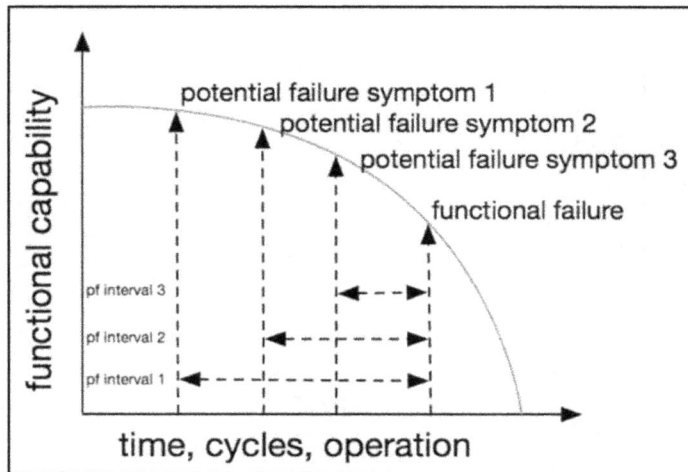

Figure 2-5. Multiple P-F intervals occur before Functional failure

In addition, Functional failure does not happen without several stages. If a person gets an illness and it's not treated, then first they may feel nauseous, then they have a fever, then they have coughing and may sweat, next thing you know it's a zombie apocalypse.

The same thing is true for component failure. There are multiple P-F intervals that can be identified as symptoms.

The key is to find these Potential failure events before the Functional failure. This will often come down to monitoring of statistical deviations.

So, if you know how your components will fail and the curve patterns they follow, then you can ensure that your monitoring intervals are within the P-F intervals.

Both environment and age factors can be considered accelerators to failure.

Environment is often overlooked, but should be taken into consideration. How an environment affects components primarily comes down to a few variables:

1. Temperature range for operation
2. Humidity
3. Power variations

Exposure (radiation, precipitation, salination, flora, fauna, air contaminants)

Components generally have defined operating limits set by their manufacturers. These are guidelines and they are often conservative in nature. Understanding the operating limits and how the use cases will lay within them is important for calculating the P-F intervals.

Any environment that is not as controlled as a datacenter needs to be evaluated against the component operating limits. By creating a table in a spreadsheet or database that has all components in an environment and the operating limits defined, you can then map the use case environment metrics to those and use them as monitoring thresholds. By doing so, you can proactively estimate environment related failures and what it will affect.

2.1.4 Graceful degradation

Graceful degradation is a concept under the realm of fault management, as opposed to fault tolerance. Meaning, it does not require extra resources to be effective. If a component has a 100% chance of failure given the right variables or acceleration factors, then a plan needs to be in place for when that happens. You can think of it as a plan B, or akin to a reserve parachute. If a component fails, will it be catastrophic? Will there be time in the P-F interval to do something? What will be impacted? Will the effects be isolated, or will they cause a cascading failure?

Fault Management is the process of finding a fault, isolating it and repairing it.

Fault Tolerance is adding extra components to a system to allow it to continue operating at full capabilities after a fault occurs.

The domino effect is the opposite of graceful degradation. When one domino piece hits another, then another, the effects are far reaching. Graceful degradation is the process to ensure that only a single piece fails at a time.

Figure 2-6. The Black Knight exhibits graceful degradation when attacked by King Arthur in Monty Pythons Holy Grail.

Graceful degradation is a design process that ensures that total failure does not happen in one fell blow. An example we can look at is the Black Knight from Monty Pythons Holy Grail.

The Black Knight lost an arm, and then kept on fighting. Then he lost another arm, then his legs and still kept fighting.

Each time, his effectiveness is reduced, but his capabilities are not entirely dismissed.

From a systems side, we could look at ECC memory. When it has a data corruption event occurring in RAM, this will often cause ill desired effect such as crashing the system. With ECC memory, it will identify the corruption and work around it to continue operations.

If we look at storage, we can use a RAID array for an example. In a RAID array, if a hard drive fails then the performance of the storage system will be reduced until it is repaired. It is still serving data but in an inferior performance state.

Graceful degradation can be applied so that the core operations of a service continue until they fail again or are repaired. How many failures a system can tolerate will determine the overall survivability of it. Aircraft, for instance, can sustain on average seven critical failures before going down.

2.1.5 Defense in Depth

Defense in Depth is a military strategy (created by the Roman army in the second century), which was used to delay an attack by using layered methods for protecting against individual forms of attack. At the time border provinces were used as multiple defense theatres, slowing down attack and allowing for more maneuvering and response.

This was in contrast to having a single fortress with walls that could be breached by a powerful enough attacking force.

This strategy was adopted by the NSA and used as a method for protecting information technology assets. The layers would consist of several of the following:

(reference https://en.wikipedia.org/wiki/Defense_in_depth_(computing)

Anti-virus software

- Authentication and password security
- Biometrics
- Demilitarized zones (DMZ)
- Data-centric security
- Firewalls (hardware or software)
- Intrusion detection systems (IDS)
- Logging and auditing
- Multi-factor authentication
- Vulnerability scanners
- Physical security (e.g. deadbolt locks)
- Timed access control
- Internet Security Awareness Training
- Virtual private network (VPN)
- Sandboxing
- Intrusion Protection System (IPS)
- Microsegmentation

This same approach can also be used in building resilient systems. Some of the layers in this respect would be:

- Fault tolerance components
- High availability software
- Ingress and Egress load balancing
- Storage multipathing
- Multiple Ethernet fabrics
- DNS based load balancing
- Availability zones
- Application level clustering
- Host based clustering

2.1.6 Survivability Analysis

In the realm of statistics, models are used to extract a certain type of data from data sets. Depending on the desired information, the data is mined using one or more of these models, then interpreted. Models have certain assumptions that help approximate the distribution of some variables based on the relationships with other variables.

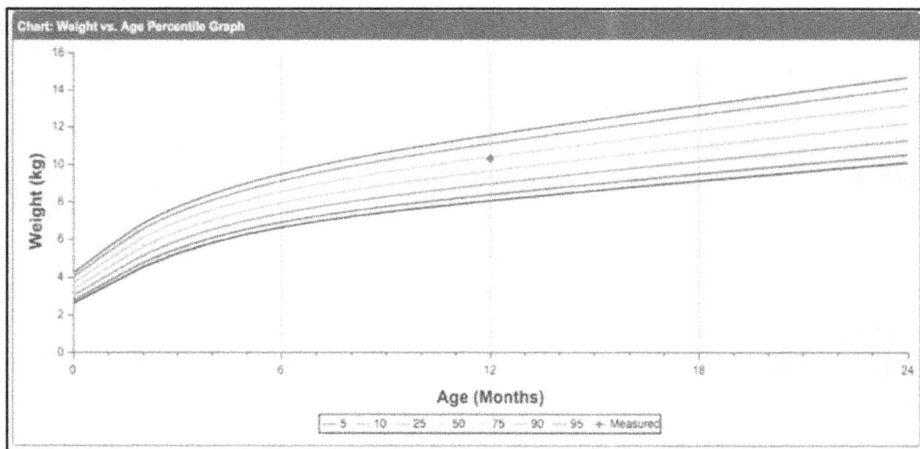

Figure 2-7. Baby growth chart showing the usefulness of model estimation.

To give an example: if we look at a baby growth chart, we can see that at 12 months the weight distribution ranges from 8kg to 11kg.

Based on the percentile range that they are in, we can extrapolate what the expected weight will be at points in time in the future. An 18month old baby that was previously in the 95th percentile at 11kg when 12 months, will likely be 12-13kg unless other variables have changed.

In biostatistics, a common model that is used is called the Kaplan Meier model. It is used to determine the likelihood of an event happening within a sample period. Although that event is often the death of a patient, we will be using it to determine the likelihood of component failure. If we take a number of components (our data set) and test them to determine their "survivability" over a period, we should be able to determine at what point they will fail based on where they fit onto the curve.

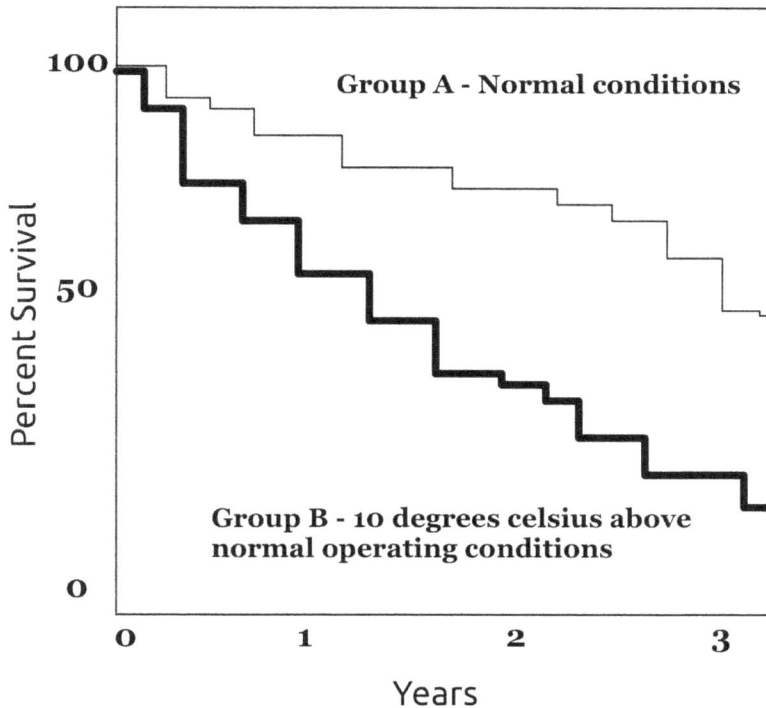

Figure 2-8. Kaplan Meier model for component survivability

The key point is that the components must fail within the sample period; otherwise they are "censored" out of the sample. This will often require a large number of components to be truly accurate.

Each component will be characterized by three variables:

1. How long they are being sampled (serial time).
2. The status at the end of the sampling (failed, or censored).
3. The study group that they are in.

During the serial time of the analysis, there is a sample rate hence the stepped look of the graph.

If we look at Figure 2-8, of note is the relationship between group A and group B. We can see that after about 1.5 years, the curves parallel. The difference between group A and group B is called the "Hazard Ratio". With one changed variable between the samples (in this case a 10-degree temperature increase), we know that at a given time period group B will have a lower survivability rate than group A based on that ratio.

If we were to take the 2-year mark as a measure, group A has about 75% survivability, whereas group B has 50%. Therefore, the hazard ratio for a 10-degree temperature increase is 1.5.

So, in conclusion, if all other variables remain the same, increasing the temperature by 10 degrees will cause 1.5 times more component failures.

Now if every component failure that ever occurs is added to this sample set (as long as they fit into the defined groups) then the accuracy of the curves will increase over time.

2.1.7 Lockstep Hardware

Applications are becoming more and more resilient by being designed for clustered environments to compensate for a single point of failure (SPOF) scenario. This is what is referred to as the scale-out architecture. If we look at a three-tier application, we have the presentation, application and database tiers. Each one of those tiers would require a clustering solution. One method is to utilize load balancers for each tier, and have data synchronization occur on the back-end. Another method is to utilize software-based application clustering that has a dedicated heartbeat between servers. If the heartbeat goes down, then the surviving server will take over the identity of the cluster and continue serving data.

These are just a few examples of how to perform scale-out application clustering. Applications that do not support clustering and are also considered mission critical, poses a serious risk.

In most servers, if a major component fails, then the server will fail. Graceful degradation can only account for so much. Examples of this are a motherboard / chipset, CPU, memory, I/O plane, etc. This in turn will cause a complete outage of the processes running on that server.

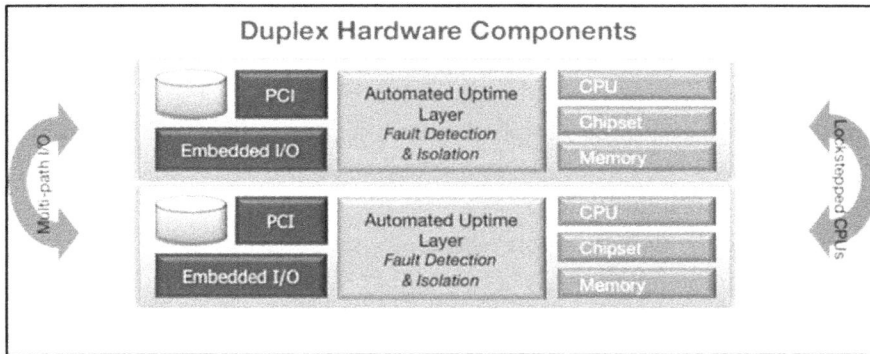

Figure 2-9. Lockstep hardware design. Courtesy of Stratus Technologies

One solution to this problem is by using lockstep hardware. This can be thought of as two identical servers in a single chassis with the exact same hardware. Each server is running all processes at the same time.

The cpu, memory, I/O are all synchronized. It is seen as a single system at all times. In the case of a failure, the surviving server will continue processing tasks with zero downtime and no performance degradation.

This solution may be more cost effective than an application clustering solution and provided a higher reliability and uptime rating. If the design mandates 99.999% uptime of greater, then lockstep hardware is definitely worth a look. Currently, Stratus Technologies is the most well-known vendor for lockstep servers, with their ftServer models.

2.1.8 Self-Healing Systems

As computing systems become more complex and more pervasive, the ability to manage them requires many sets of skills and many support groups.

If you look at a large organization, you will see silos of knowledge; network teams, infrastructure teams, service desk, desktop management teams, storage teams, Linux/Unix admins, Windows admins, application admins, security, etc. The list goes on and on.

All those departments can be summed up in a few key categories:

1. Test / Dev
2. Provisioning

3. Operations
4. Maintenance

Or if you want to put it in even more simplistic terms, let's use a light switch scenario:

1. People who try new light switches.
2. People who turn the light on upon request.
3. People who keep the lights on.
4. People who make sure the lights will turn on every time.

How people perform their job role can often be defined by a flowchart (as seen in figure-18). If a logic loop is programmed into the system itself to perform some of the tasks that are currently performed by staff in one of those four identified categories, then it can automate some of the maintenance and repair process.

The problem with that theory is that the programming of the system to automate those functions needs to be less costly than the employment of those individuals that do the job. In addition, other roles need to be added within the organization to monitor the functioning of the system automation and make modifications when necessary.

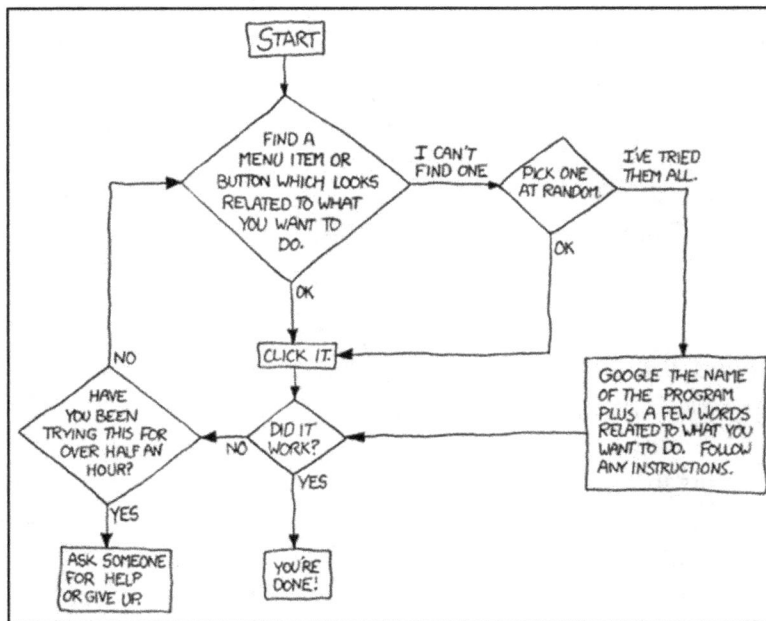

Figure 2-30 Common service desk flowchart. (as per xkcd.com)

A compromise is to automate some of the low hanging fruit by performing very simple actions very quickly. One way to do this is with a monitoring system that has ties into a scripting engine, or orchestration repository. Then using situational identifiers and thresholds, pre-defined remediation processes can occur.

An example of this is VMware vRealize Operations and vRealize Orchestrator. Based on certain parameters defined in vROPs, a vRO workflow can be instantiated and do any number of complex actions on the environment. It is one of the most powerful platforms in the industry for doing that.

If you want to dive a bit deeper into automation, you can start reading up on Autonomic Computing and the initiatives associated with that.

(Reference: https://en.wikipedia.org/wiki/Autonomic_c33eomputing)

The main concepts of an autonomous system are these:

1. Self-configuration: Automatic configuration of components.
2. Self-healing: Automatic discovery, and correction of faults.
3. Self-optimization: Automatic monitoring and control of resources to ensure the optimal functioning with respect to the defined requirements.
4. Self-protection: Proactive identification and protection from arbitrary attacks.

Each one of these concepts may require several systems in place to realize. However, it is the devising of the framework for these that is the hardest part. This is not a product; this is an approach. The approach will involve many products, duct-tape, patience, coffee, alcohol and a glue gun.

2.2 Availability

2.2.1 MTTR

MTTR is also known as Mean Time to Restoration or Repair, or in simple terms, "downtime". The length of the downtime depends on a few factors. I have divided it into sections A, B, C, D:

A. Fault Detection Time (MFDT)

- How long before the outage is noticed.

B. Preparation and or Delay

- How long before the appropriate parties are made aware of the issue.
- Length of time for planning logistics.
- How long before a plan is formulated to address the issue.

C. Mean Active Repair Time (MART)

- How long to come to agreement on the solution.
- How long to get the appropriate parties to remediate the situation.
- How long to verify the solution.

D. Administrative Delay

- How long to notify all affected parties and stakeholders that corrective action has been performed and that the services are back online.

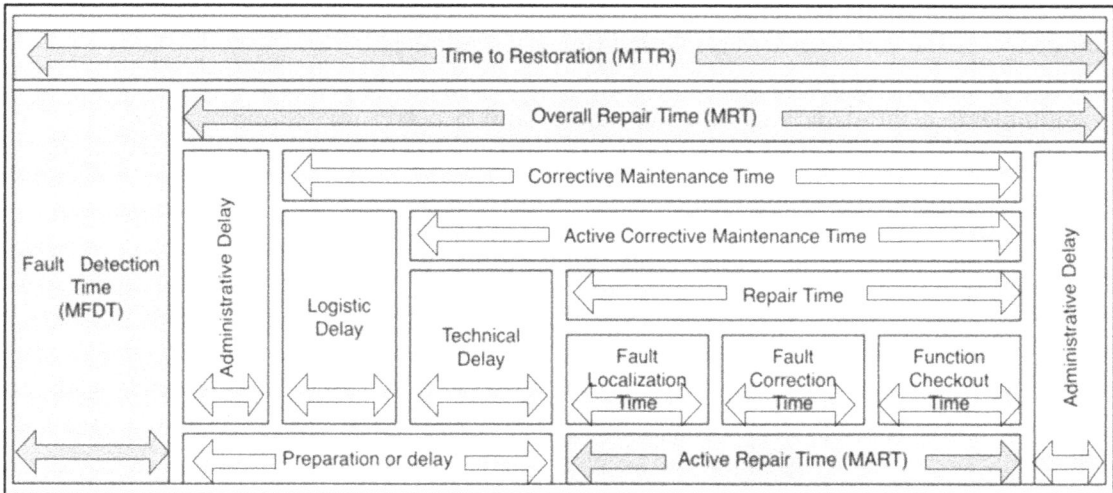

Figure 2-10. Repair time taxonomy

After these items are performed, then the system that was affected can now continue operations as per normal. This does not mean that the work is over, but that services are functioning again. Other things that will have to be done behind the scenes are a root cause analysis, a post-mortem impact review and an action plan for preventing it from occurring again in the future.

The overall repair time (MRT) is a combination of the areas B, C, D. The downtime (MTTR) is MFDT (detection time) + MRT (repair time).

This is a great way of providing visibility into the work done to remediate the issue, but it is not reflective of the actual risk that needs to be taken into account during the process.

What assets and services does the outage affect? Will those "other" services and assets that interact with this one be put at risk during the repair process?

I'll give an example of this.

A business has a memory-based database that writes to a high-speed cache, then to spinning disk every few minutes. This database is key to the core business and downtime will cost millions of dollars every minute as well as loss of customer confidence and financial SLA penalties for being down.

The datacenter has two channels for power, A and B, backed by independent battery banks and generators.

A maintenance event failure in the datacenter takes out power channel B. The datacenter has identified the issue and estimated resolution time is 60 minutes.

All equipment in the datacenter is using both A and B power for redundancy, except for the network equipment between the high-speed write cache and the spinning disk. It only has a single power supply which is on power channel B.

This means that although all services are up and running, every second that the database is not writing to disk is a potential loss of data. If power-channel A goes down, all database transactions that occurred since power channel B going down, will be completely lost and irrecoverable. In addition, the write cache is not meant to store writes for long periods of time. If it hits the maximum cache size or duration, it will halt all further operations freezing the entire database.

The goal in this case, would not be to repair the issue, but rather to remove the risk of loss of data. Moving the offline network equipment to channel A would be the quickest way to do that. Failing that, changing the database to "read-only" would provide some level of functionality but prevent data loss.

This is what is referred to as getting to a "safe state" during repair.

Sometimes, getting to a safe state elongates the overall MRT, but it reduces the time that the business is at risk.

An organization may also set limits on the maximum time that is permitted for downtime. At that point, the business has fail-safes to ensure that if the ideal strategy does work within a defined time period, the less-ideal plans start coming into play. Frequently, they are heavy handed, and although they invariably work, there is a massive cost to them.

An example would be if a person stepped on a nail and got an infection on their foot. Then perhaps they have poor circulation and can't cycle through enough clean blood, so necrosis starts killing the tissue around the infection. If treatment for the infection did not work within a defined period of time, then the doctors would have to amputate to prevent the infection from spreading to the rest of the body.

In IT, an example would be if a mail server was down and did not come up within a defined period of time, then DNS records could be changed so that all mail went through a third party cloud provider.

This resolves the immediate risk of losing mail, however it causes a huge swath of problems with users that need access to mail, scheduling, historical archival, etc. Then since the operational effort would be to assist the users experience, less effort would be directed towards the actual problem, thus elongating the MTTR.

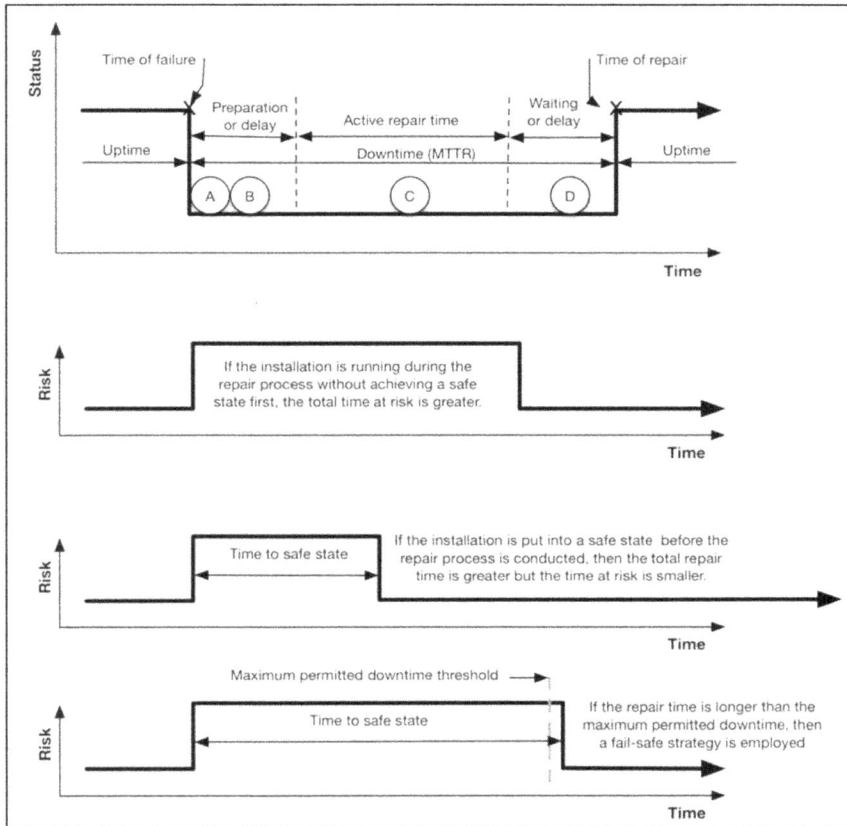

Figure 2-11 Risk during a failure

2.2.2 MTBF and MTTF

Mean Time Between Failures or MTBF, is also referred to as "uptime". It's not quite as simple as that though. It is the average measured time between failures. It refers to systems that are repairable. If a system is not repairable, then the formula for MTTF (Mean Time to Failure) is used. If a system fails and is repaired a number of times then fails permanently, it will then have both MTBF and MTTF which will be different.

If you have four identical systems and ran each one until they permanently failed, then the MTTF would be the average amount of time that they were operational.

You can have an MTTF for a single system, without others in the equation, but that's more of an individual lifespan vs an average.

If we called these systems W, X, Y, Z then we could say that the systems failed as per below:

System W: 8554 hours

System X: 6774 hours

System Y: 9012 hours

System Z: 7995 hours

The MTTF in this case is the average of the four data points, which is 8083.75 hrs

Now let's look how that would look like if we looked at the MTBF for System

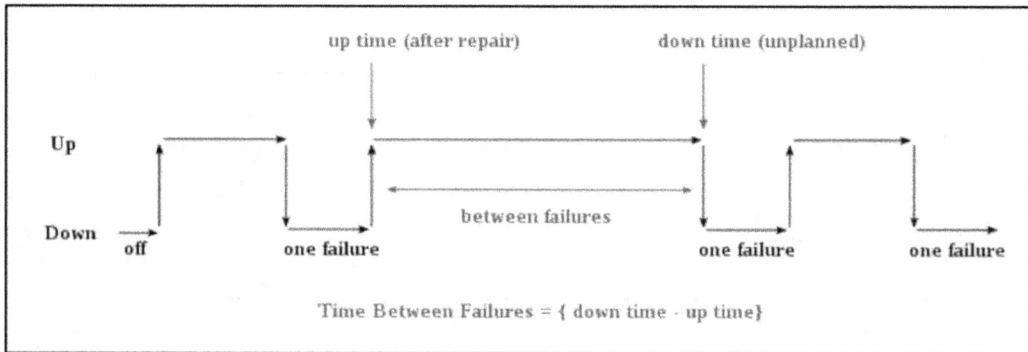

Figure 2-12 Time between failures

System W	Failure period	Time between Failure (TBF)
Failure 1	672hrs – 676hrs	668 hrs
Failure 2	1344hrs – 1348hrs	1161 hrs
Failure 3	2509hrs – 2513hrs	838 hrs
Failure 4	3351hrs – 3355hrs	3048 hrs
Failure 5	6403hrs – 6407hrs	127 hrs
Failure 6	6534hrs – 6538hrs	1216 hrs
Failure 7	7754hrs – 7558hrs	Combined TBFs: 7058 hrs

The MTBF is the total of the combined TBFs / the number of failures over the lifetime of the system. So, in this case it is 7058 / 6, which is 1176.33

This means that based on the historical trends with all variables being the same and identical systems, there is a high probability that the system will suffer a failure approximately every 1176 hrs (a month and a half). Each failure causes approximately 4 hrs in downtime (MTTR). It will also have an average total lifespan (MTTF) of about 8083 hrs (11 months).

These values for System W, are made up, but knowing the real values of your environment allows you to accurately estimate the availability based on projected failures. In this scenario, the availability is 99.65%, or 28 hours of downtime over 11 months.

There is more than one type of MTBF metric, if you really drill down to the details. Other types are:

MTBCF:
Mean Time Between Critical Failures

MTBUR:
Mean Time Between Unscheduled Removal

MTBSA:
Mean Time Between System Aborts

If a system is not repaired, then the metric will change to MTTF (Mean Time to Failure)

If a failure needs to be differentiated from a non-dangerous situation to a dangerous one, then MTTFd will be used. An example with a critical multi-site active/passive system would be a site-failure in the DR site (MTTF), vs a site-failure in the primary site (MTTFd)

2.2.3 Five Nines

The calculation for availability is something is that is often advertised by service providers, datacenters, and other organizations as a major selling point. What this is really all about is perception. Customers expect they should be able to access services 24/7 from anywhere in the world. If you want to ensure that your services are available all the time to customers, then you will want to make sure that your service provider does the same for you.

Avail%	Downtime / yr	Downtime / mo	Downtime / wk
90% (1x nine)	36.5 days	72 hours	16.8 hours
99% (2x nines)	3.65 days	7.2 hours	1.68 hours
99.9% (3x nines)	8.76 hours	43.2 mins	10.1 mins
99.99% (4x nines)	52.56 mins	4.32 mins	1.01 mins
99.999% (5x nines)	5.26 mins	25.9 seconds	6.05 seconds
55.5555555% (9x fives)	162.2 days	13.52 days	3.15 days

The definition of what is classified as downtime is where some service providers get tricky. For instance, to some service providers scheduled maintenance is not considered downtime, even though services are offline. Another trick is not measuring when only a subset of users is affected. To them this may also not be considered downtime. Performance SLAs should also be tied to Availability SLAs. Just because something is online does not mean it is usable for the required function.

So be wary when service providers tout five nines, as it may be nine fives.

2.2.4 Service Level Agreements

Service level agreements are created between service providers and customers. In these, will be the details of what is measureable and in-scope for the desired service. The language in these can be deceiving, so great care must be taken when reviewing them. If an organization is providing a service to customers and is using a service provider to do so, then it cannot offer a greater uptime than the underlying provider. If they do, for example say 99.9% uptime for its clients, and the underlying service provider states 99.99% uptime (excluding maintenance and events affecting less than 90% of users), then they may not be able to deliver on the agreement.

Sometimes, an SLA breach will cause cascading affects to service provider clients, and their clients and their clients and so forth. If a service provider fails to deliver on an SLA, then there may be a monetary fine that needs to be paid. Below is an example:

Priority	Penalty Factor	Remediation
Critical	1/30th total monthly revenue	$500 / day
High	1/30th device/system monthly revenue	$7 / day
Medium	1/30th device/system monthly revenue	$7 / day
Low	1/30th device/system monthly revenue	$7 / day

2.2.5 Service Level Objectives

Service level objectives are the measurable components in an SLA, such as bandwidth, latency, jitter, etc. Key attributes in an SLO are as follows. An SLO must be:

1. Attainable
2. Measurable
3. Repeatable
4. Understandable

An SLO is usually for a defined period of time, such as a month.

Often organizations will have service level objectives internally. In this fashion, they treat the company users as clients and define the SLO for services rendered. Doing so keeps the innovation and management of technology within a company strong. It also assists in promoting good communication and accountability between departments.

2.3 Redundancy

Redundancy is when there is more than one component or system that performs the same function. This comes into play when the criticality of the system is high and the dependent services are of higher than normal importance. By having multiples of the resources needed, the overall availability is increased. One example from earlier in the book is lockstep hardware. However, there are many ways that redundancy can be perceived, not just in duplicate hardware. I'll dive into some of these below:2.3.1 Redundant Components (internal)

This goes back to lockstep hardware, which I will start with. This provides higher availability for critical systems that may experience a hardware failure and have no application aware redundancy. Good for single application critical systems with legacy coding.

2.3.1 Redundant Components (external)

This is essentially doubling the hardware for critical systems. This provides higher availability by creating mirrors of the original systems. This method more than doubles the cost for these systems because simply adding in a copy won't make it automatically work in tandem. Load balancing techniques need to be employed as well. How the mirroring of components is done depends on a deeper look at the applications that are the core services being served up. One example is a three-tier web application, with a front-end presentation layer, an application or "logic" tier and a data tier.

Below we see the logical representation of non-redundant vs. redundant from a very high level.

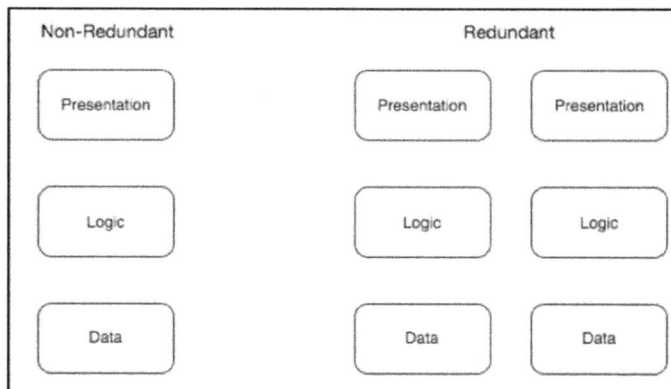

Figure 2-13 Redundant Components

The problem with this representation is that is does not include any of the supporting hardware required for the redundancy. This is a key concern for the design because it adds more cost, complexity, management and risk. If we provide some more insight to the level of redundancy, we can see that some additional equipment has been added.

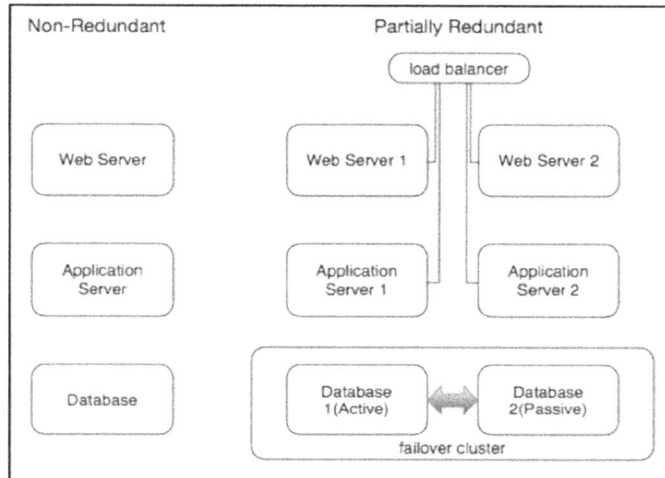

Figure 2-14 Greater redundancy in components

A load balancer has been added to allow for failover of incoming connections to the web servers and application servers.

This does not work for the database servers, and thus a failover cluster with active and passive servers is created.

This is still a bit simplistic, as it does not account for the SPOF (single point of failure) of the load balancer, or the network equipment.

What started out as simply "double the components", does not truly provide the desired outcome. A failure of the load balancer would now render four servers unavailable. Or worse, if the underlying network were not composed of redundant hardware, then all six servers would be offline.

To take this a step further, what if the uplink to the rest of the world had an issue. That would be another SPOF. Taking all these variables into consideration let's have a look at a "more redundant" network for this three-tier application.

The diagram in figure 2-15 is a better representation of external hardware redundancy on a single network.

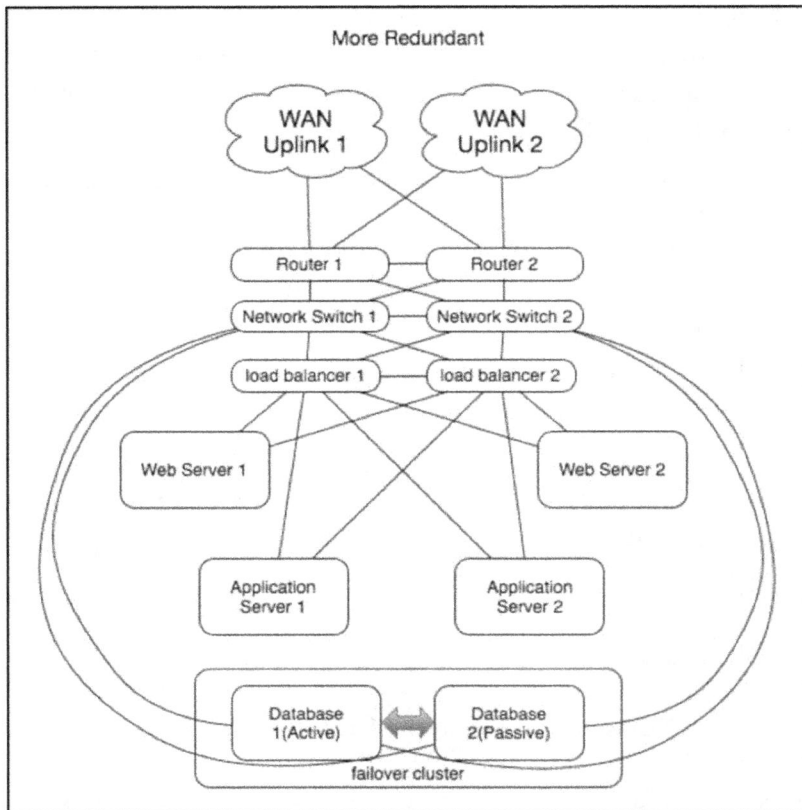

Figure 2-15 Even more redundancy built into the design

This is still not a fully redundant application and there are still risks involved. Some other considerations are the possibility of power failure, a site failure, DNS failure, or a regional power failure.

Applications that are cloud-native are much easier to architect in a manner that allow for use of microservices and hyperscaling. This reduces the burden of application based clustering (which may have scaling issues) and puts more focus on load balancing systems.

The "fully redundant" design is very difficult to achieve. It is essentially the technological unicorn for infrastructure design. It can be talked about at length, but is rare to truly see in the world.

In order to achieve this, the end-to-end communication has to be considered. This also includes the client, which can be difficult to control in many circumstances.

2.3.2 Redundant Resources

This would include personnel that administer the environment, as a loss of them may create a large risk in the case of a failure that is not self-healing. Hours of coverage and after-hours support and escalation come into this category. People also need vacations, so any role that does not have one or more backup personnel for that role is a risk. Some organizations will have a buddy system with two people minimal per technical role. Some organizations rely on the general competency of their staff and put the risk in the hands of the supporting documentation. Other organizations mandate cross training across technical staff to avoid tribal knowledge and silos of operation.

2.3.3 Passive Redundancy

Sometimes redundancy can be built into a system without adding additional components. Rather, the existing components can be bolstered, or overdesigned. In engineering, this is called the safety factor. In infrastructure design, it is used when there is an unknown variable that may cause the environment to surpass its design thresholds.

The safety factor can be quantified with the following equation:

$$\text{Safety Factor} = \frac{\text{Failure Load}}{\text{Design Load}}$$

Depending on the criticality of the component or system, a safety factor of 3 to 10 times may be applied. This allows systems to compensate for times of stress, misuse and failure. What the safety factor number itself applies to in a design depends on the area that is being bolstered. It could be heat failure threshold, it could be processing power, it could be storage capacity, or it could be component reliability associated with price. It really depends.

2.3.4 Active Redundancy

In the early 1980's, the standard large-scale ship computer for the US Navy and Submarines was the IBM AN/UYK-43. This was one of the first systems that employed Active Redundancy.

Active Redundancy performs three key actions automatically that previously would have had to be done manually:

1. Fault Detection

2. Fault Isolation

3. Reconfiguration

Because of these actions, systems with failures like CPUs, RAM, I/O boards, etc. will continue to operate by logically removing the failed components from the system.

This concept requires a fault based if/then/else logic, which can be considered a simplistic artificial intelligence.

Other systems that commonly use Active Redundancy are satellite systems, aircraft and the electrical power grid.

2.3.5 Data Redundancy

There are many ways to have redundancy in data, here are a few:

1. RAID Arrays – Use of multiple redundant disks and parity data to ensure data integrity
2. SAN Mirroring – Continuous real-time replication of data to a secondary storage array
3. Replication Factors – By replicating data to several places, data integrity is more achievable
4. Storage Array level snapshots – These snapshots store an index of all data on the storage at a specific time. The index maps the block locations to actual files, which allows for instantaneous restores.

2.4 Business Continuity

Business Continuity is a set of strategies for ensuring that regardless of the event that occurs, there is a plan for restoring business operations. These strategies are comprised of four main areas:

1. Resilience
2. Recovery
3. Avoidance
4. Contingency

2.4.1 Resiliency

By using concepts such as active and passive redundancy, defense in depth and application clustering, critical services can continue to operate despite a failure event.

2.4.2 Recovery

A comprehensive plan is created to restore data and systems in the case of data loss, corruption, compromise or a failure event.

2.4.3 Avoidance

Disaster Avoidance (DA) uses technologies such as stretched layer 2, metro cluster storage, and automated recovery. It must be used when SLAs with five nines are a requirement.

2.4.4 Contingency

Plans are made to respond to unforeseen or unavoidable events. A contingency plan is normally deemed a last resort, in case that the resilience and recovery strategies did not work as desired.

Lack of business continuity plans within an organization, will virtually guarantee that in the face of a critical failure event, the organization will go bankrupt.

Often the Business Continuity Management methodology is referred to the BC/DR plan. These strategies should be very well documented, with the specific roles and associated people called out within the documentation. Anybody should be able to read the documentation from the beginning to end and fully understand the process that is required to restore the environment to its original state. According the FEMA, 40% of companies do not reopen after a disaster, and another 25% fail within the first year afterwards. The Small Business Administration reports 90% of small businesses fail within 2 years of a disaster. That is why you need strong contingency plans.

2.5 Regionalization

If you carve up the earth into areas that can most easily service the populous with WAN connectivity and computing power, you end up with regions. Regions do not entirely equate to political borders, but some do match up with some continents (as can be seen from the image below).

By having compute resources in a specific region, you can more effectively service clients that require services.

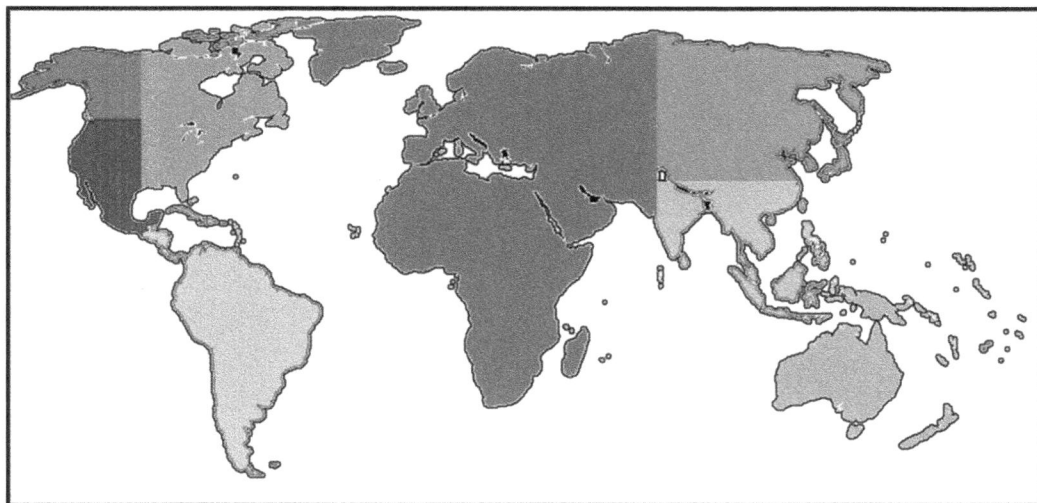

Figure 2-16 World map split into availability regions

2.6 Availability Zones

This term became popular by Amazon Web Services, which made it a core part of its strategy. An availability zone is essentially a data center within a region. By spreading out compute and storage across availability zones, the risk to the organization caused by a site failure is reduced. Often time availability zones will also be intra-regionally diversified as well inter-regionally diversified. This protects against region and single site failures; however, some additional considerations need to be taken into account, such as inter-regional latency.

2.7 Infrastructure Diversity

On the path to an optimal environment state, an organization will go through a number of policy changes. One of these is often the preferred vendor for hardware. It may make it easier to go for a single vendor stack, such as Cisco, HP, Dell, IBM or Huawei. You have a simplified product portfolio to choose from, so procurement and design is easy. Having a single vendor infrastructure stack reduces the total cost of ownership.

Now let's look at it from the opposite perspective. A single vendor often has a single unified code base across multiple pieces of equipment. If an exploit is found for one piece of equipment, then it is also probable that other pieces of equipment are also vulnerable by the same exploit.

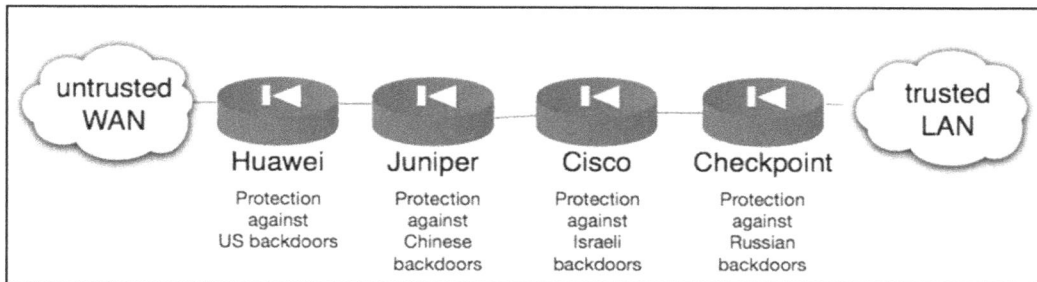

Figure 2-17 Protection from government sanctioned backdoors.

Then we have to think about government-sanctioned backdoors to the equipment. Depending on where the equipment originates, there may be exploitable backdoors put in the system that allow unmonitored access from third parties.

Sometimes the backdoors are simply using a type of encryption that is easily crackable by a certain organization. Other times it may be that there are some 0-day exploits that an organization knows about but has not publicized.

Then there is the discussion around best of breed. No single vendor dominates in every category that they have equipment in. There will always be smaller upstart vendors that do something better than the big players. This is the way the market operates, and will never change. So why not get the best equipment for the specific requirements? Get firewalls from vendor A, switches from vendor B, IPS and perimeter security from vendor C. This will ensure that you always have the top of the market for each area of your environment.

In 2016, the FBI filed a lawsuit to get backdoor access to the Apple IOS Operating System. Apple refused as it was a violation of user privacy and trust. Soon after, the FBI dropped the lawsuit. However, the FBI had already hacked the Apple iPhone in order to access information that may have been on a suspect's' phone. The FBI publicized that they did indeed hack it, but they refused to share the method of the security breach with Apple. Apple was not complicit in this instance, but what about other hardware vendors instances?

This proves that the practice of hoarding exploit is real and that likely, the majority of government sanctioned exploit kits are not made known to the vendors until they are no longer useful.

In 2017, a massive data breach at the NSA, exposed a treasure trove of non-published 0-day exploits used by the agency to bypass the security of Operating Systems, software and hardware around the world.

The risk with this is the requisite knowledge set. Since there are so many different vendor architectures, management stacks and areas of focus, you will require personnel that can effectively be classified as experts in each of these areas. Unless there is an organizational policy to cross train technical staff, additional people with the required expertise will need to be hired. The next issue that will occur from this is that when you have more staff you will need to have the management structure in place to accommodate them. If the staff are not managed effectively, then you will have gaps and overlapping of job duties. This inefficiency will create a greater total cost of ownership.

Now let's talk about security in-practice as it applies to infrastructure diversity. If you were to look at the percentage of security breaches by root cause, you would see that on average, exploitation by firmware defect is about 1/20 the amount compared to exploitation by misconfiguration.

So, let's think about this for a second. If you have a team with a strong skillset on hardware that is from a single vendor, versus a hodgepodge of skillsets on best of breed hardware, then you are less likely to encounter a security breach or system failure from misconfiguration. However, if the staff has been trained to a minimum qualified level, sufficient for day-to-day management and advanced administration, then the best of breed hardware is a better option.

2.8 Application Resiliency

The resiliency of an application is can be measured by MTTR. Often there are many components and services that plug into an application. How the application responds to failures of those individual components and services is critical to the uptime. Some applications may have queues, retries or wait periods while other operations are being called. This continual polling process can cause latency in the entire system and impact the user experience of all users, especially in a shared environment.

One solution to this is pre-loading the environment with a generated load to see what the effects are, based on a load of a certain number of users. Then take these effects and use them as indicators in a monitoring system. The monitoring system will trip on the threshold and remedial action can then be manually or automatically performed.

Another solution is to make use of a fail-fast system with a circuit breaker library. These systems will continually monitor the application internally as opposed to externally and it will determine the maximum threshold for a wait operation, then fail it immediately and provide a detailed error to the end user and perform remedial operations. On a periodic timeline, the system will do a single retry operation in the background to check if the system is back up, then allow requests to continue as per normal.

An example of this in action is Netflix Hystrix. It is a library that monitors latency of services and fails it quickly if it is not performing at the minimum required level. This allows for a user experience that is consistent and without the possibility of a cascading failure.

One drawback of this approach is that it can only be used when the application makes use of micro services. Applications using a monolithic architecture will need to use a different approach.

2.9 Application Clustering and Failover

Application clustering requires a few key elements:

1. At least two hosts of similar capabilities
2. Shared storage or database
3. A heartbeat network in between the hosts

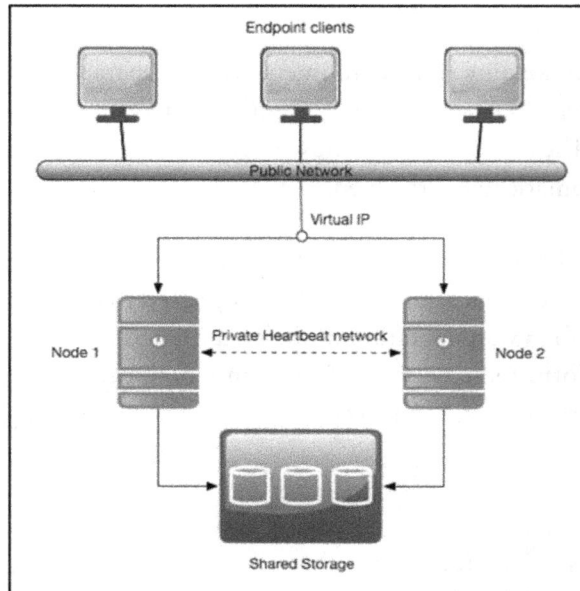

Figure 2-18 Application clustering model

In the two-node failover cluster represented in Figure 2-18, there is an active and a passive server. Each server has a unique IP address, but there is also a "cluster address", which is a virtual IP. The active server will own the cluster IP, and all traffic will direct to it.

In the case of a failure, the heartbeat status will show as failed and the passive server will be promoted to active, and thus assume the additional cluster virtual IP.

This is the simplest clustering technique and it is quite effective. When a client connects to the cluster IP, it will be for a specific service, such as CIFs, a custom application, database, etc. Only the defined cluster services will run associated with the cluster IP and not all services on the server will be made available, even if they are running on both servers.

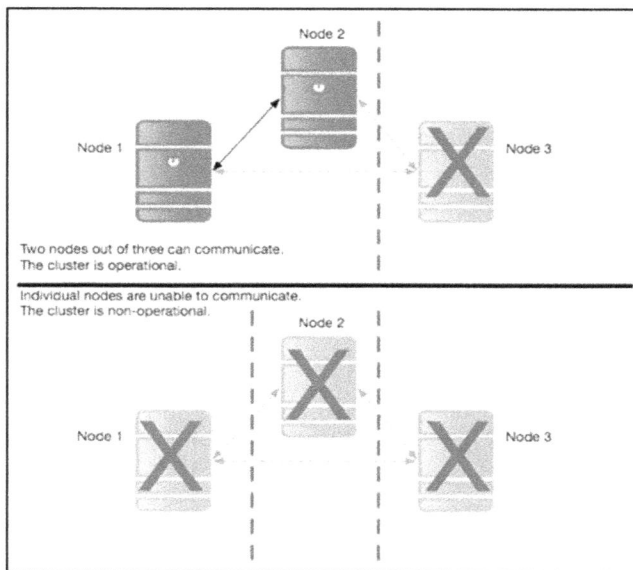

Figure 2-19 Cluster node communication

When more servers are required due to load, etc., then often load balancers are utilized. In that scenario, a cluster IP and resources are also defined, but there is no active / passive hierarchy. There is simply pools and states. A pool is a group of servers running the same service, where all servers are active.

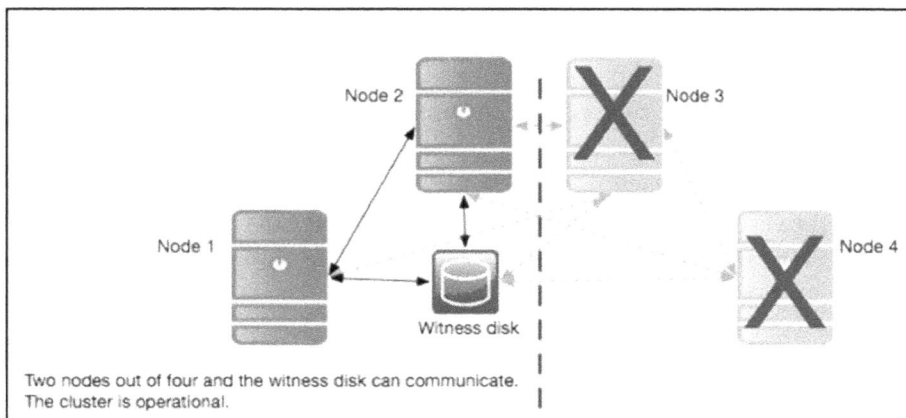

Figure 2-20 Witness Availability

The load-balancing algorithm may depend on network load, server load, number of connections, etc. The strategy depends on service and what works to provide the best overall experience for the end user.

If the cluster shared service goes offline on a specific server in a pool, then it will be disabled or ejected from the active pool.

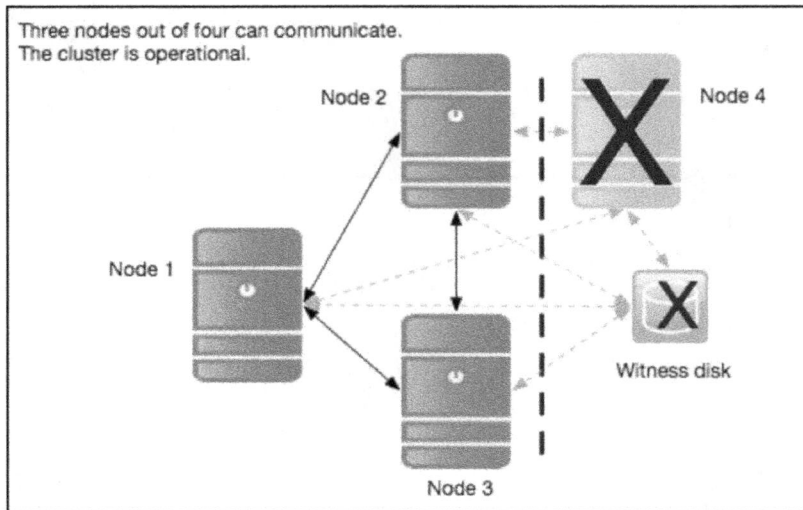

Three nodes out of four can communicate.
The cluster is operational.

Node 2

Node 4

Node 1

Witness disk

Node 3

Figure 2-21 Node and Witness Unavailable

Another clustering technique makes use of a quorum. A quorum dictates how many failures that a cluster can endure before it is offline. In Figure 02-21, we see that on the top, one server is offline, and the cluster still runs.

In Figure 02-22, we see that inter-server communication is not functioning properly.

Even though the services on the servers may be running, the cluster is offline.

Servers in a cluster may also make use of a witness disk, which contains a copy of the cluster configuration. In the case where there are groups of servers that can communicate with each-other, but not with all servers, the majority group will stay online and the minority group will go offline.

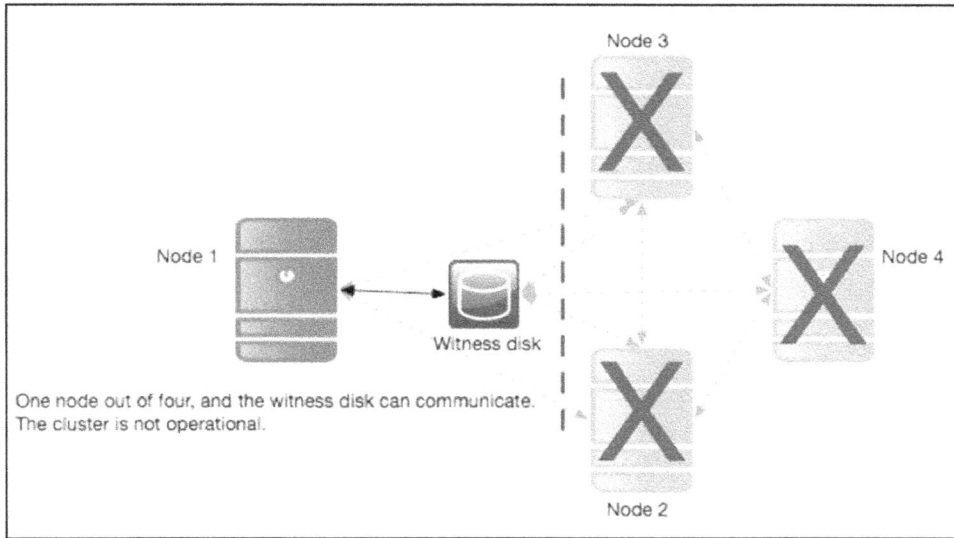

Figure 2-22 Cluster witness availability

This happens due to a voting system. Each server will cast a vote and the majority of votes will win the cluster ownership. When network communication resumes, then the servers that were offline will start serving their resources again in the cluster.

vMotion has been the single most beautiful feature that has ever been introduced to the VMware ecosystem. Some say that not only was it the catalyst for the success of the company, but also for the entire industry. The first application that was tested successfully during a vMotion was the pinball application on Windows.

2.10 Host Based Clustering

When the resources of a host (Storage, Memory and Compute) are abstracted by a hypervisor, they can be added to a pool of resources from other hosts. This pool of resources can then be made available for virtual machines to consume.

A group of hosts in a cluster also have other advantages such as the ability to migrate active compute workloads from one host to another automatically (VMware vMotion and DRS for example), or restarting a virtual machine on another host (VMware High Availability). With hosts within a cluster that meet the requirements (primarily network bandwidth), real time memory replication can be accomplished (VMware vSphere Fault Tolerance), allowing for the failure of a host with no effect on the applications running on the protected VMs. This mitigates the risk of downtime for critical legacy applications that do not support other application clustering techniques.

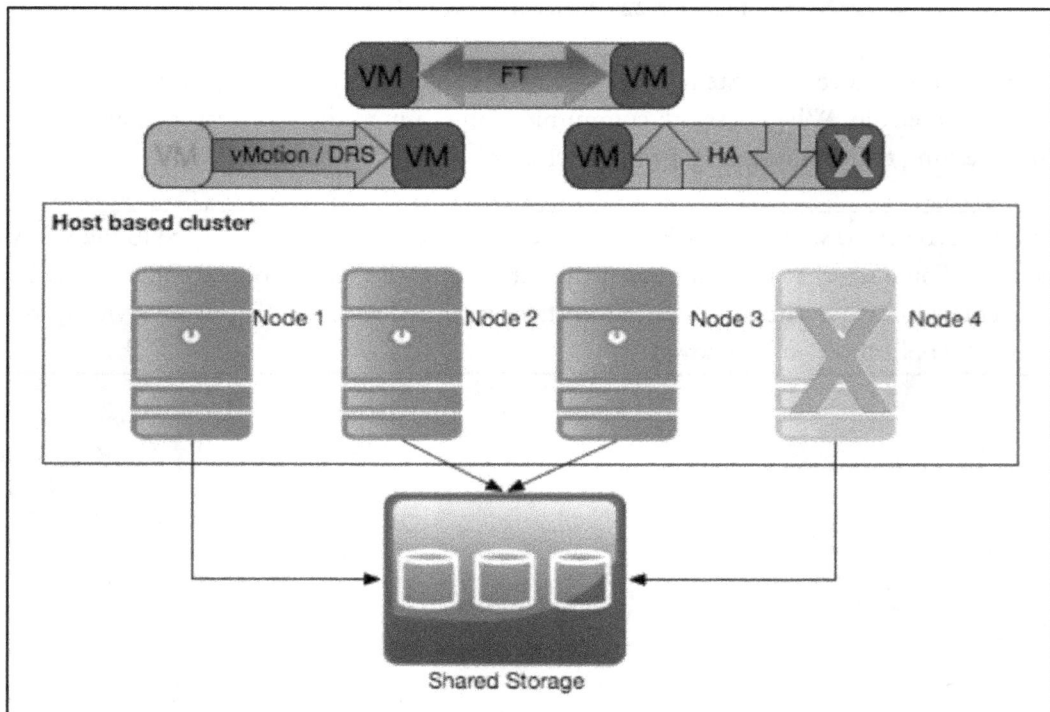

Figure 2-23 Host based clustering

2.11 Content Delivery Networks

When users are connecting to a SaaS application, or a website that has a lot of users from many different regions, then content delivery networks or CDNs come into play. They have a regional copy of the data that is closer to the end user. Less distance is fewer hops, lower latency, and better performance. Some CDNs will completely host the content and applications in each region and perform web content optimization (WCO) to ensure the best end user experience.

This strategy allows a CDN to employ additional security measure such as make use of a Web Application Firewall (WAF) or auto scale the CDN nodes to absorb a DDOS attack. Akamai and Cloudflare are both examples of providers of these services.

2.12 Load Balancing

When a client connects to a server, it does so via IP or DNS. The more client request come in, the more resources are required of that server. There are two possible solutions for this; scale-up or scale-out.

Figure 2-24. Chief Brody is hitting a resource limit.

The best way to describe a scale-up solution is to remember the movie JAWS. There was a scene when they were trying to catch the shark and it ends up ripping the boat apart because they were not prepared for it.

At that point, Sheriff Brody said," We're gonna need a bigger boat!"

To accommodate increased load, a scale-up solution will add more RAM, CPU, network bandwidth, or storage to a single node. This will work up to the point where the chassis or unit is unable to accommodate it. Then the model of unit will likely be increased to a larger version. The cost of this type of solution is predictable in a stepped manner. There will be a large capital cost with a larger performance gain, then a slow incremental cost until the capacity is at its threshold or the lifespan of the hardware is at its end.

One of the main issues with a scale-up solution is that it's often the single point of failure. To combat this, double or triple the hardware is purchased in order to allow for an active/passive failover, or a standby system.

A scale-out solution makes use of many nodes that work in parallel, but are presented as a single system. Scale-out can be done at many different points in an infrastructure; such as with clustered storage, network switch stacks, web services, application services and database services.

The most important part of a scale-out architecture is to be presented as a single entity, which is performed by a load balancer. A load balancer will present a single IP address to the network with a service on it. The inbound connections will then be redirected to a pool of resources in the backend.

A good example of this is with web services. As the number of connections increases, the CPU, memory and network load increase on a server. To alleviate this, simply add another identical server to the pool and they will equally share the load (depending on the algorithm). As load increases or decreases, the resources required can be added or removed from the pool.

There are two main categories of load balancers; those that act on the transport layer (layer 4 on the OSI model) such as IP, TCP, UDP, and those that act on the application layer (layer 7 on the OSI model) such as http headers, cookies, or the value of a parameter in the message.

Servers in the pool will be monitored for health, load, number of connections, etc. to ensure that they are still fit for processing requests. If they are not, then they are removed from the pool.

The last type of load balancer is called a global server load balancer, or more commonly known as DNS (though some may argue that a GSLB is more than just DNS).

The DNS standard allows for round robin load balancing of records. This means that if you created two A records with different IPs for mcboatyface.designingrisk.com, that the DNS servers would automatically cycle through the IPs every request. This is a very simplistic load balancing mechanism.

However, if one of those IPs is unreachable, then 50% of the requests will go to that unreachable IP. A GSLB has more intelligence because it will add, remove or prioritize IPs from the record based on availability, load, etc.

2.13 Micro Services and Cloud Native Applications

A micro services architecture is composed of many services doing a single function. They are easy to initiate and easy to replace. Micro services are normally built on top of a container platform, such as Docker or VMware Photon Machine. The containers are then controlled by a container manager such as Kubernetes or VMware Photon Controller.

The container manager will ensure that the overall health of each instance is maintained. If an instance is not healthy, it will be terminated and respawned immediately. Through the container manager, applications can be instantiated and scaled as required.

Cloud Native Applications are built from the ground up for the cloud (pun intended) to be highly scalable and elastic. They are the next evolution of software architecture incorporating many concepts discussed in this chapter, such as scale-out, resiliency and fail-fast methodology.

2.14 Enterprise Architecture Document Design Workflow

Every design has a series of documents with it that define what the problem is, the guidance for the solution and the justification for the design. How these documents are created often depends on the internal structure of the organization.

Some organizations provide controls that guide the solution, which are based on established and tested methods and processes. An example of these are NERC CIP (North American Electric Reliability Corporation - Critical Infrastructure Protection), PCI DSS (Payment Card Industry Data Security Standard) or NIST (National Institute of Standards and Technology). Depending on the industry, the source may be a combination of controls from several sources, or a subset of only one.

Some organizations simply define the scope of the project and take known physical designs and map them to the problem. This is often seen with VARs that have a small number of vendors they deal with and limited number of architects. The number of solutions sold is more important to them than the "fit" of the solution. It leads to designs that are quickly out of date, more expensive, or underspecified.

In keeping with the design inputs and outputs philosophy using CARRD and the enterprise architecture design method, the following document workflow is defined. It contains ten documents for the design phase of a project.

These documents may not be the only ones that will be created and there might be some debate on what the scope of the Architecture Document Set includes. Other documents in this category are the "Build Guide", "Operations Guide", "Runbook" and "Testing Methodology Guide".

These documents are often created in parallel with the set of 10 documents shown, however they may also be created exclusively in the Deploy and Operate phases.

The order shown in the flowchart in Figure 02-25 insinuates a waterfall funnel model. A document is created, then acts as an input for the next document, then a number of documents input into the end deliverables which are the design documents. This is mostly true, but as you progress through each document there may be discoveries which will change some of the former documents.

This is no truer than when you create the Risk Analysis document. This document could cause changes in the non-functional requirements document (NFRD) and the Dependency Mapping Document. It also acts as an input to the Design Decisions document and therefore affects the three design documents.

1	Scope of Work	A scope of work document, which outlines the assets and resources which will be included, as well as the overall goals of the project.
2	BRD	A business requirements document (BRD), which outlines the underlying business objectives. This document will include some requirements as well as some assumptions to provide direction for the solution where it may have not existed otherwise.
3	FRD	The functional requirements document (FRD), which provides guidance for what the solution must be able to accomplish from a technical standpoint.
4	NFRD	The non-functional requirements document (NFRD), which outlines the constraints for the project. Often this is included in a single document with the FRD, called the FNFRD
5	Risk analysis	The risk analysis document, which outlines all current and future risks. It can be used for justification of design decisions, or in a current state analysis as justification for an infrastructure change.
6	Dependencies	The dependency mapping, which shows what components are associated with other components. This helps determine risk and aids in obtaining resources for change management, BC/DR, and maintenance.
7	Design Decisions	The design decisions document, which lays out the thought process for the following designs. It borrows justification and reasoning from all the input documents
8	Conceptual Design	The conceptual design document, which provides a high-level overview of the solution.
9	Logical Design	The logical design, which provides a vendor agnostic framework for the solution. The logical design does not expire unless there are drastic architectural changes to the infrastructure
10	Physical Design	The physical design, which provides low level guidance for hardware models, firmware, etc. This aids in the deployment process, however it will expire when a new firmware version or hardware model is released.

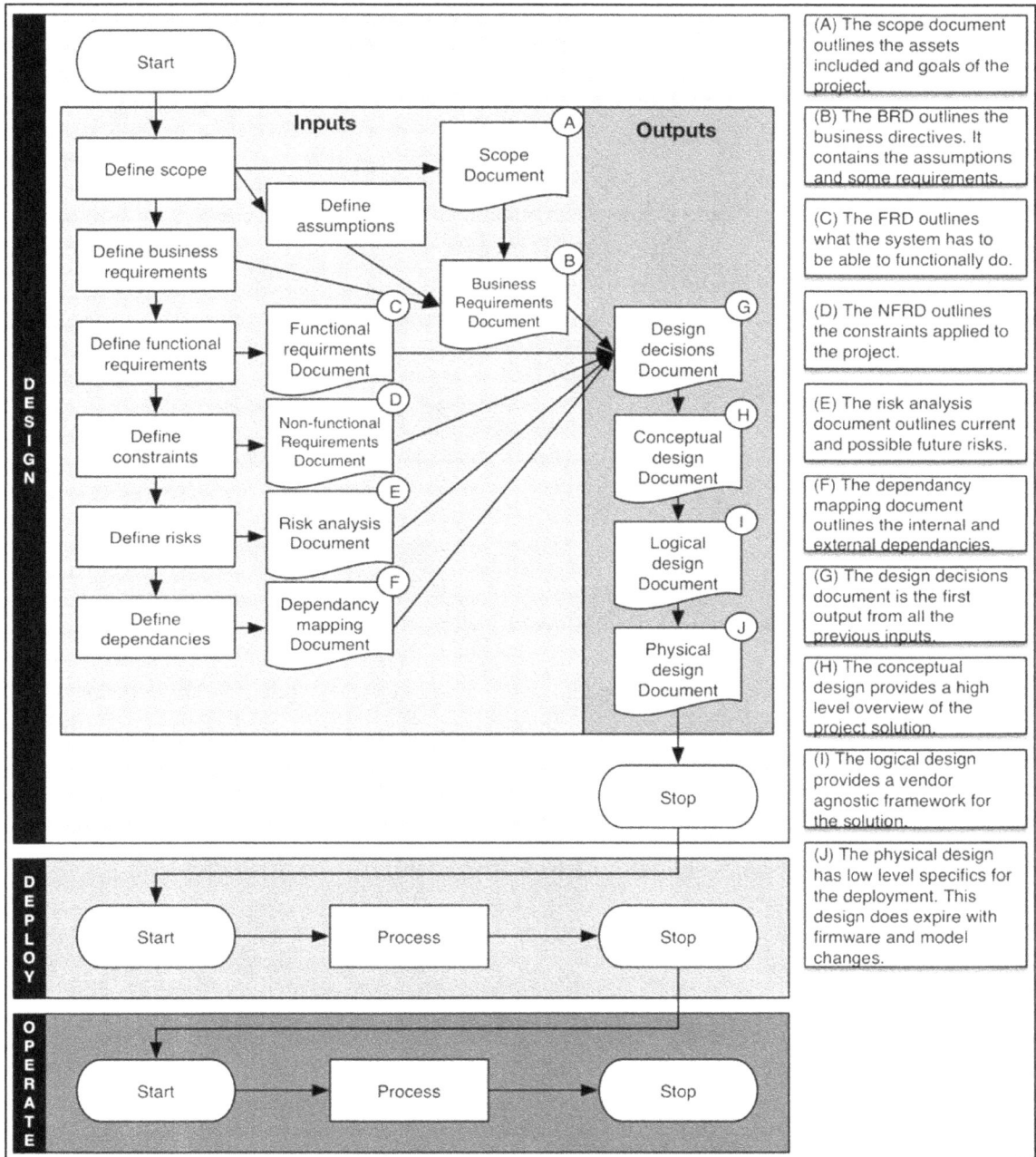

Figure 2-25 Document Workflow

The diagram contains the following annotations:

(A) The scope document outlines the assets included and goals of the project.

(B) The BRD outlines the business directives. It contains the assumptions and some requirements.

(C) The FRD outlines what the system has to be able to functionally do.

(D) The NFRD outlines the constraints applied to the project.

(E) The risk analysis document outlines current and possible future risks.

(F) The dependancy mapping document outlines the internal and external dependancies.

(G) The design decisions document is the first output from all the previous inputs.

(H) The conceptual design provides a high level overview of the project solution.

(I) The logical design provides a vendor agnostic framework for the solution.

(J) The physical design has low level specifics for the deployment. This design does expire with firmware and model changes.

2.15 Indestructible Reference Architecture Example

This reference architecture example illustrates possible manners in which the topics of discussion in this chapter can be put to use. This is not a design document, but simply several perspectives of the same architecture from different levels. The point here is to understand that each diagram, although shown independently, is interconnected with the others in a way to show maximum resiliency.

The first diagram below demonstrates at a high level, the network connectivity of the solution. There are three regions each with four availability zones. Each availability zone represents a data center.

Zone A and B for each region are physical data centers managed by different companies. However, every Zone A (in each region) is the same company, as is Zone B, etc. Zone C is a cloud services provider, as is Zone D. Each one of those is again, a different company.

Zone A and B each have 3 WAN providers, with mesh networks. Each WAN provider has a peering agreement with the other two. All physical data centers use these links to communicate with one another.

Within every region, each physical data center can communicate with each cloud service provider via a VPN or overlay network, such as VXLAN.

The cloud services providers can communicate with the same zones in other regions via the intra-cloud network. The separate cloud providers can also communicate with each other via VPN / VXLAN within each region.

Every cloud services availability zone uses the internet connectivity provided by that company. Every physical data center availability zone will use the internet connectivity provided by each of the WAN providers, for a total of three uplinks each.

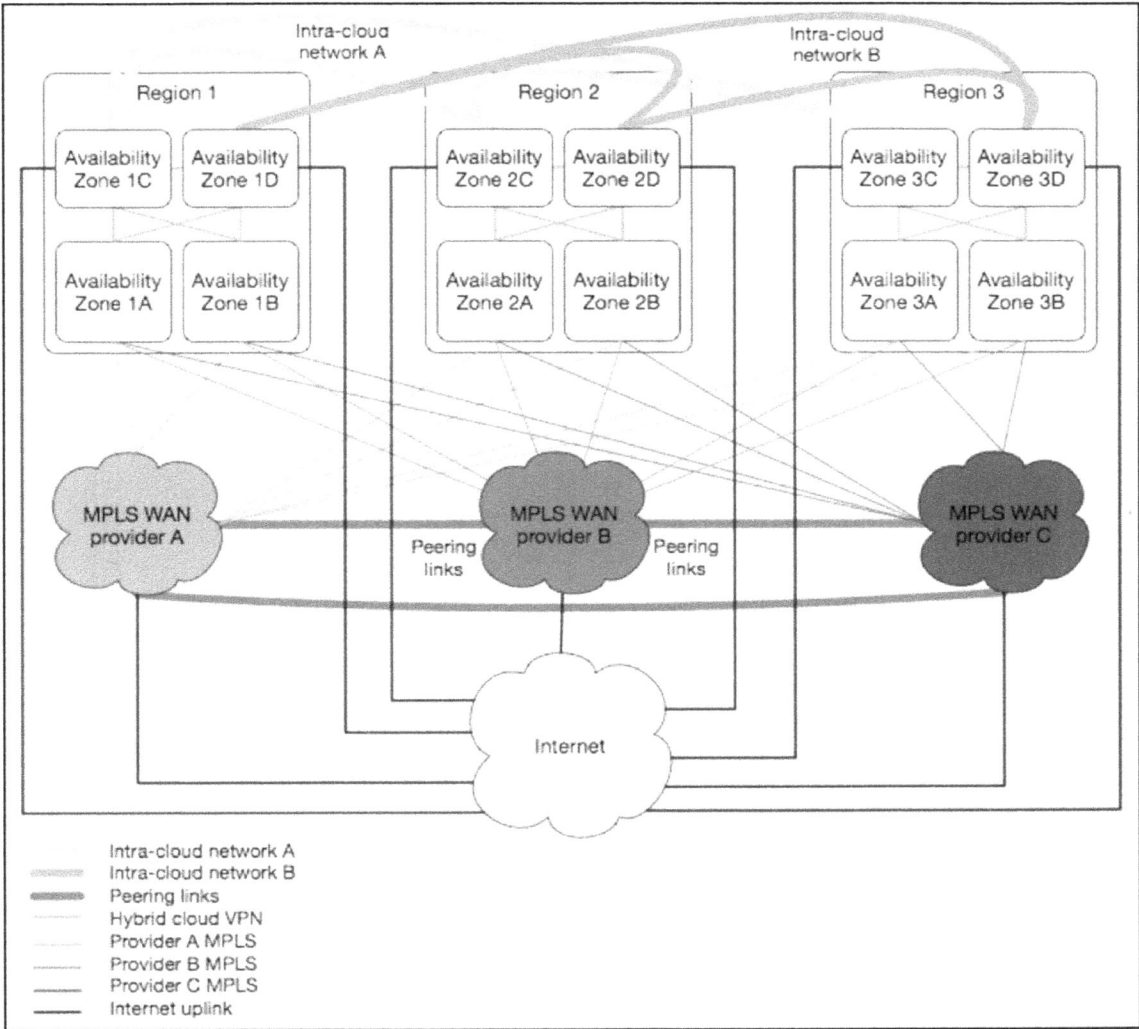

Figure 2-26 Indestructible infrastructure high level diagram

The image below shows a single region of the previous diagram

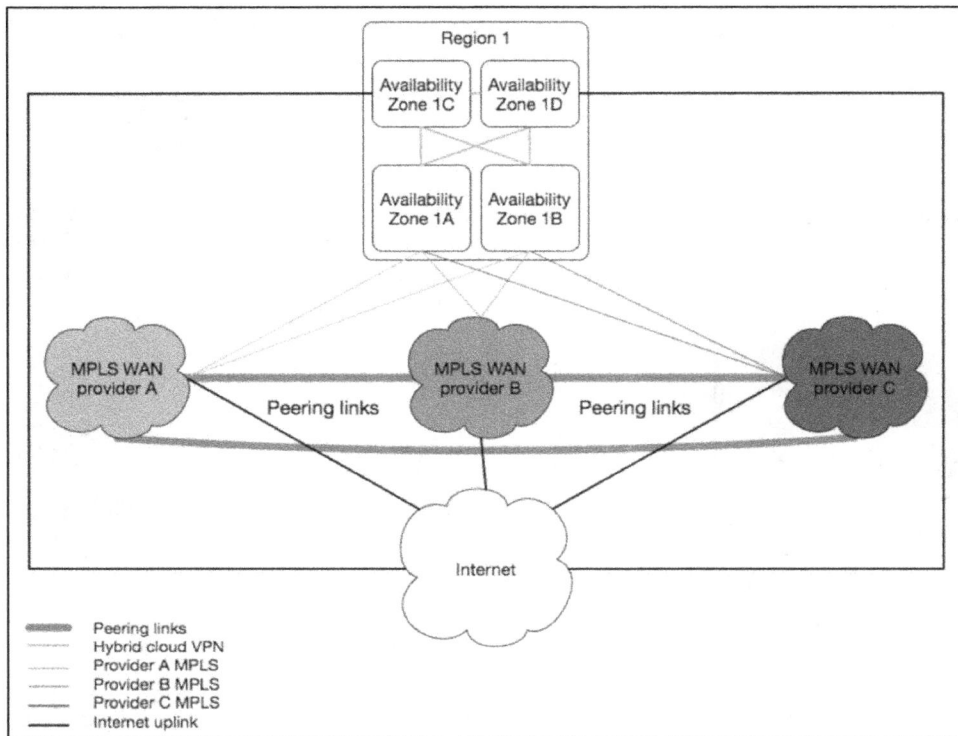

Figure 2-27 Indestructible infrastructure high level single region diagram

The diagram in figure 2-28 illustrates some of the components comprising each availability zone.

The physical availability zones (on-premises - identified by the number 1) have dual active routers, redundant top of rack switch stacks, highly available load balancers, traditional storage nodes, traditional compute nodes and hyperconverged storage/compute nodes. There are dual power channels for each rack, backed by two separate UPS banks. Each UPS bank is backed by 3 high powered diesel generators (not depicted, but implied by general good design).

The cloud services providers (identified by the number 2) utilize an Infrastructure as a Service model, which means that the provider manages the underlying infrastructure and maintains an SLA ensuring availability. All of the components depicted in the cloud are also running on the core infrastructure in the physical data centers.

The applications provided are layered on top of the components in each one of the availability zones. There are a combination of legacy enterprise applications and cloud native applications that are served.

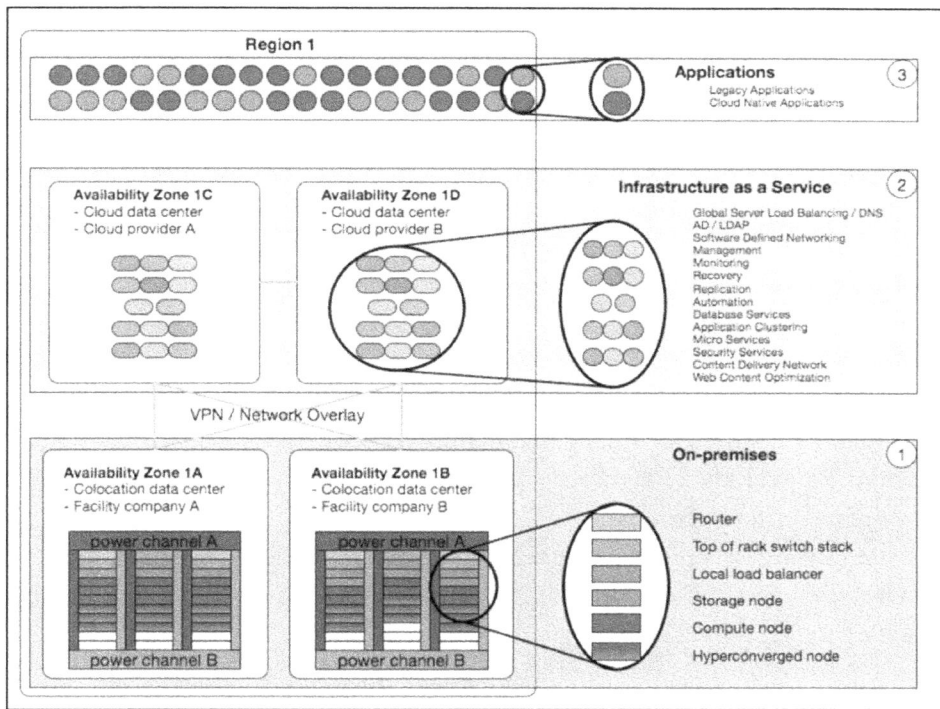

Figure 2-28 Indestructible infrastructure single region logical diagram

Technologies such as replication, global and local load balancing, monitoring and automation ensure that; data is consistent, applications are available, responsive and that there is allowance for multiple concurrent failures in every region without impact to users.

The diagram below depicts technologies that act within a single region, vs globally.

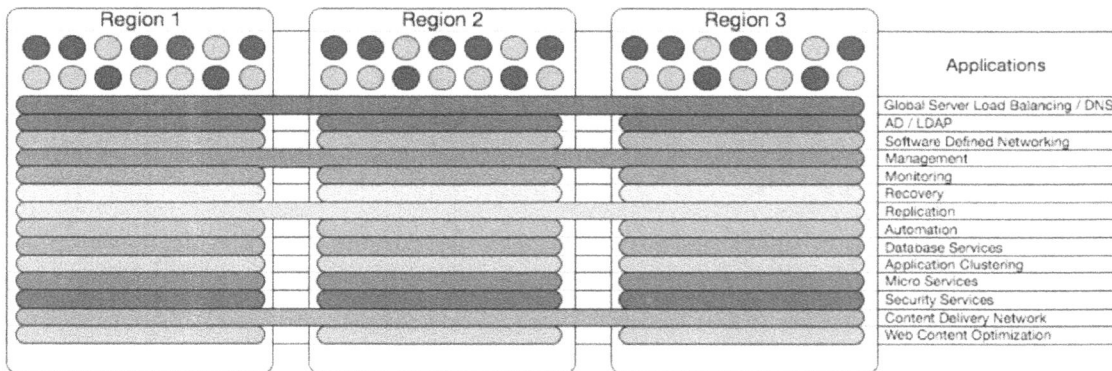

Figure 2-29 Indestructible infrastructure multi-region logical diagram

2.16 Chapter Summary

1. A **reliable** component is continually validated, whereas a **dependable** component has been validated at some point and there has not been enough data provided to counter that yet.
2. When a component is operated at stress levels that are higher than normal, it is called physical acceleration.
3. By performing a stress test, or burn-in test of a solution, you are more likely to find components that will fail early and replace them.
4. The period from initial symptom to total failure is called the P-F interval. The P-F stands for "Potential Failure" to "Functional Failure".
5. Graceful degradation is the process to ensure that only a single piece fails at a time and the system is still operational.
6. Defense in depth is a protective method that uses multiple layers of resiliency.
7. Survivability Analysis allows you to determine the average lifespan of a particular component.
8. Lockstep hardware uses mirrored hardware within a single chassis.
9. Self-healing systems employ introspection and automation systems to resolve issues that would normally be addressed by a human administrator.
10. MTTR is also known as Mean Time to Restoration or Repair, or in simple terms, *downtime*.
11. During MTTR, it is important to get to a *safe state* quickly.
12. Availability is measured in terms of "nines". 99.9% uptime is 8.76hrs downtime per year. 99.999% uptime is 5.26mins downtime per year.
13. Service level agreements are created between service providers and customers. They often have details that if not accounted for, can cause considerable risk.
14. Service Level Objectives are the measurable components of an SLA.
15. There are two type of component redundancy; internal and external.
16. People need to be considered a resource and redundancy is required, as are recovery and replacement plans.
17. Passive redundancy uses a safety factor, which is equal to the failure load / the design load.
18. Active redundancy utilizes an embedded monitoring and response system that logically removes failed hardware.
19. Data redundancy can include RAID arrays, SAN mirroring, replication and backup technologies.
20. Business Continuity is a set of strategies for ensuring that regardless of the event that occurs, there is a plan for restoring business operations.

21. A contingency plan is normally deemed a last resort, in case the resilience and recovery strategies did not work as desired. It is often heavy handed.
22. If you carve up the earth into areas that can most easily service the populous with WAN connectivity and computing power, you end up with regions.
23. An availability zone is a data center within a region.
24. Having a single vendor infrastructure stack reduces the total cost of ownership.
25. A single vendor often has a single unified code base across multiple pieces of equipment. So, if an exploit is found for one piece of equipment, then it is also probable that other pieces of equipment are also vulnerable by the same exploit.
26. A fail-fast system with a circuit breaker library will continually monitor the application internally as opposed to externally and it will determine the maximum threshold for a wait operation, then fail it immediately and provide a detailed error to the end user and perform remedial operations.
27. Application clustering requires at least two hosts of similar capabilities, shared storage or database heartbeat network in between the hosts.
28. Host based clustering provides a more comprehensive, yet simpler approach to availability than application clustering, but it cannot provide 100% availability during multiple concurrent failures.

2.17 Chapter Review Questions

1. How do you calculate the total MTTR for an outage? What is the approximate percentage of active repair time compared to the total outage time?
2. What is the largest average cause of age related failure in components? What about non-age-related failure?
3. What business continuity strategy protects against users accidentally deleting data?
4. What business continuity strategy protects against large scale national disasters?
5. If you had a legacy application that ran as a single service, how could you ensure the highest level of availability?

CHAPTER 3

Know what you don't know

"Noise is relative to the silence preceding it. The more absolute
the hush, the more shocking the thunderclap."
(Quotation from the graphic novel "V for Vendetta")
--Alan Moore (Writer)

Success may destroy a company, but many small failures may make it stronger.
This chapter deals with ideas that are counterintuitive or often unspoken.

> "By three methods we may learn wisdom: First, by reflection, which is noblest; Second, by imitation, which is easiest; and third by experience, which is the bitterest."
>
> -Confucius

Learning through personal experience can be bittersweet. On one hand, it provides a very real perspective that you internalize and won't forget. On the other hand, the greatest opportunities for learning are often during the times of greatest crisis or failure.

In this chapter, we will discuss the learnings from the failures of others, so that in your future endeavors, history does not repeat itself.

3.1 TCO for five nines

Do you have "Five Nines" of availability?

Figure 3-1. In German, the word "nein" means "no".

"Five nines" of availability has become a buzz-phrase these days, and also a box that many organizations want to check off when referring to the availability of its own infrastructure.

Management loves it and it also looks good on marketing material, but what is the true cost of ownership for five nines? Is it worth it to be able to guarantee? Or is it something that is expected but impossible to deliver?

If it is something that is mandated, then let's see what needs to be considered in order to provide it. Below are items of consideration that need to be addressed. Each item has the following properties: name, reasoning / justification, whether they are preventative or reactive, and the level of cost associated with them.

1) **SLA payout cost**

Reasoning:
If you provide an SLA to clients, what is the agreed cost of a breach?

Preventive / Reactive:
Reactive

Cost:
Moderate

2) Insurance for failures

Reasoning:
If there claims against your company due to IT failures, you can be protected by obtaining IT errors and omissions insurance.

Preventive / Reactive:
Preventive

Cost:
low

3) Legal fees

Reasoning:
This includes fees such as a having a lawyer on retainer, and the cost of litigation itself.

Preventive / Reactive:
Both

Cost:
high

4) Costs of legal actions from clients

Reasoning:
These are any damages that would be paid to clients if legal proceedings are brought forth.

Preventive / Reactive:
Reactive

Cost:

5) Vendor SLA

Reasoning:
This is the support agreement with software and hardware vendors that ensures that if there is a failure, that there will be 24/7 support to resolution.

Preventive / Reactive:
Preventive

Cost:
moderate

6) Service provider SLA

Reasoning:
This is the agreement with upstream services providers that dictates the monetary penalties and recourse for failures they have.

Preventive / Reactive:
Preventive

Cost:
low

7) Operational response times

Reasoning:
This cost is comprised of staff being available 24/7 to deal with issues.

Preventive / Reactive:
Preventive

Cost:
moderate

8) Failure tests, spot checks

Reasoning:
This is the cost of time and effort to perform tests to ensure that the environment responds as expected when forced into a controlled failure state.

Preventive / Reactive:
Preventive

Cost:
low

9) IT Governance

Reasoning:
The costs associated with ensuring that IT objectives align with business objectives.

Preventive / Reactive:
Preventive

Cost:
moderate

10) Emergency response teams

Reasoning:
Companies need to ensure that there is a team in place to respond as quickly as possible to failures. There is a preparation and training cost, then the time and materials for reaction.

Preventive / Reactive:
Both

Cost:
moderate

11) Change management

Reasoning:
There is a cost to pay for the resources responsible for setting up the change management practice and operating it.

Preventive / Reactive:
Preventive

Cost:

12) Public relations

Reasoning:
There is a cost for maintaining a public relations team, or retaining a PR company. Then there is the cost for "Spin Doctoring" a failure to minimize public perceptive damages.

Preventive / Reactive:
Both

Cost:
low

13) Technical staff coverage

Reasoning:
In order to ensure that operations are running in a highly available manner, a minimum number of staff is required. In order to keep it that way, additional persons are required to take some of the workload of the primary staff.

Preventive / Reactive:
Preventive

Cost:
low

14) Sufficiently hardware redundancy

Reasoning:
To reliably maintain high availability for the environment, a minimum level of hardware redundancy needs to be employed.

Preventive / Reactive:
Preventive

Cost:
moderate

15) Sufficient software redundancy

Reasoning:
To reliably maintain high availability for the environment, a minimum level of software redundancy needs to be employed.

Preventive / Reactive:
Preventive

Cost:
moderate

3.2 Determining the cost of failure

When something fails in an environment, whether it's hardware, software, process, or people, there will be a cost associated with the impact. Sometimes the cost is simply the time it takes to replace the failed component, whereas other times it may have a direct cost to the business.

The cost associated with failure also depends on when it happens, whether it is in the development, proof of concept, QA, UAT or production phase. Depending on how budgets are allocated, it may make sense to conduct more thorough testing in the pre-production phases to ensure that once in production other areas are not affected by a failure that could have been found previously.

Pre-production phases of an infrastructure often come out of budgets for initiatives, whereas production often comes out of the operations budget. Not all organizations have this delineation when it comes to IT, as it may just be an IT budget. That is often the case with smaller organization's or ones that have a less formal structure for budget allocation within a department.

This applies direct to risk since the impact to business will be different depending on when the failure occurs, it should be taken into consideration when determining if a risk is tolerable for an organizations appetite. As the phases of the infrastructure change, the risks need to be re-evaluated as well.

Here are some "soft costs" that need to be considered:

3.2.1 Permanent failure disposal

When a component is deemed unfit for use in production, lab or in scrap for component extraction, it will need to be disposed of. This cost may be paid to electronic recycling agencies, or waste removal. Depending on the usability of the scrap, it may be sellable on the bulk market to refurbishing companies, or privately on eBay, craigslist or another online forum.

An example of this is a blade server chassis that only supports old PIII blades. The cost of operation in power exceeds the value or usefulness that could be gained from it. A single high-end laptop could have more capabilities than the entire legacy chassis. The components are also vendor specific and out of support, so they cannot be used in any other manner.

3.2.2 Rework maintenance

When a component is removed from production, but still has usefulness or value if repaired or modified, then it is reworked.

An example of this is a storage array that is no-longer on the HCL (hardware compatibility list) for the production environment, but it would have value in test / Dev as an additional storage tier. The rework maintenance cost would be the time and effort to wipe the equipment, upgrade the firmware to the latest version and integrate it into the test / dev environment.

3.2.3 Rework validation

When a component is modified in the rework process, it needs to be validated before it can be used with certainty. The rework validation cost would be the time and effort to perform regression testing on the component and the environment it will be reintegrated with.

An example of this, (following on the previous example) is when a reworked storage array is put into a test / dev environment. It will need to have a full test and validation process performed on it to ensure that it is "fit for use". Changes to the infrastructure design and components used effectively make it a new addition.

3.2.4 Additional material procurement

When a component is undergoing rework maintenance, there may be additional parts that needs to be replaced or augmented to make effective use of it.

An example of this is, (following on the previous example) is when a reworked storage array is put into a test / Dev environment. If it was a fiber channel array, then the environment may need an FC switch, fiber patch cables, SFP transceivers and HBA cards in the hosts.

3.2.5 Vendor rework

If a failure is attributed to a fault in the component manufacture, or a software flaw, then the vendor is responsible for providing a replacement that resolves this. However, this is often a shared responsibility with the operations team. This cost is the time and effort to perform the directed work with the vendor to achieve resolution.

An example of this is when a modular switch has a corrupt file system. The vendor may only supply a logic board with a new firmware on it. The operations team needs to remove the old logic board, swap it for the new one, upgrade the firmware to the previous version, and import the switch configuration.

3.2.6 Component repair

When a component has a failure, there are costs associated with time and effort to perform the repair. There are also costs with replacing failed components. If a component is under warranty, then the time and effort to go through the RMA process is considered. If a component is not under warranty, then the time and effort to source the replacement is considered. The procurement cost of the replacement hardware/software is also factored in. If the replacement component is not identical, then due diligence is required to ensure hardware compatibility and there is a need to perform integration validation testing.

An example of this is if a switch fails and it was EOL (end of life), the vendor may have a new model that functionally does the same thing as the old one, but is different. In most cases the components would be similar enough that the new model will easily supersede the old with the original configuration. However, if the vendor's new model has a different code base, it may not be able to import the previous configuration. Or perhaps the vendor ended that line of products and a new vendor needs to be used. Either way, it needs to be treated as a new component and the time and effort for design augmentation, risk validation, testing, integration and possibly even training has to be considered.

3.2.7 Management time

This is the cost of time for management to provide oversight to the failure resolution process. Tasks consisting of meetings with the operations team, interaction with stakeholders and performing of an FMEA (failure mode effect analysis.)

A root cause analysis (RCA) and a failure mode effect analysis (FMEA) are complementary processes. An RCA is used to determine the reason something failed and an FMEA is used to analyze the effects of the failure. An RCA is a one-time process after a failure, whereas an FMEA is cumulative and can be performed before and / or after the failure event.

Function	Potential Failure Mode	Potential Effects of Failure	S	Potential Causes of Failure	O	Current Process Controls	D	RPN	CRIT	Suggested Actions	Ownership and target date for completion	Action Results					
												Action taken	S	O	D	RPN	CRIT
Black and white office printer	Does not print	Client is upset / Support involved	8	Out of paper/ink	5	Resource alert	10	400	40								
				Paper jam	3	Jam alert	10	300	30								
	Prints extra pages	Wasted ink and paper	6	Paper sticks together	2	Loading procedures	7	84	12								
	Takes too long to print	Client annoyed. Loss of time	3	Full queue	7	None	10	210	21								
				Large print job	3	None	10	90	9								

Figure 3-2. Example of an FMEA

RPN = Risk Priority Number
S = Severity
O = Occurrence
D = Detection
CRIT = Criticality
S x O = Criticality
S x O x D = RPN

If several failures have the same RPN value, the ones with a higher severity value take priority.

3.3 The opportunity cost of quality

First a little refresher on what opportunity cost is. Opportunity cost is the amount of something you need to give up in order to have something else. This can be presented in a couple of ways, the subjective way and the quantified way.

If you want to represent it subjectively, then you just need to understand the correlation of two things. Here is an example of a correlation:

As the price of an infrastructure component goes up, the quality of it increases. (This is generally thought to be true, although there are always exceptions.)

However, without data to back this up, how can it be determined to be true? Let's quantify that with some metrics.

The first thing we need to define is what do we mean by "quality". It is somewhat intangible, but can be measured in multiple ways such as performance, reliability, resilience or other properties.

In this instance, we will measure quality as it relates to component failure rates.

There are numerous factors that contribute to uptimes such as the quality of components, the quality of design, and the quality of implementation. You could include the quality of continual operations, but that's dynamic and can't be measured at a single point in time.

A component failure is measured by its MTTF (mean time to failure) or the failure rate. The MTTF will provide the average time before something fails, whereas the failure rate shows the percentage of failures for a specific time range. We looked at MTTF in the previous chapter, so let's look at failure rate now.

In this example, we will look at the average failure rate of hard drives over time, in one year intervals for a period of three years.

In a sample of hard drives, there are three manufacturers with 2TB hard drives. The table below shows the percentage of failures out of 100 for each year. If you were to calculate the average rate of failure over a three-year sampling period, then you get the mean failure rate.

Vendor	Size	1st YFR	2nd YFR	3rd YFR	MFR
Vendor X	2TB	29%	66%	85%	60%
Vendor Y	2TB	23%	23%	26%	24%
Vendor Z	2TB	10%	9%	11%	10%

YFR = year failure rate
MFR = mean failure rate

Vendor drive failure rates

The mean failure rate would then be used inversely as the quality index (QI) on a scale of 0-10, 0 being the poorest and 10 being without failure.

Since we are mapping the mean failure rate inversely, 100% failure is a QI of 0 and 0% failure is a QI of 10. The sample set is mapped as per below.

Vendor	Size	Mean failure rate	Quality Index (QI)
Vendor X	2TB	60%	4
Vendor Y	2TB	24%	7.6
Vendor Z	2TB	10%	9

Vendor drive quality index

Below you can see this mapping represented visually.

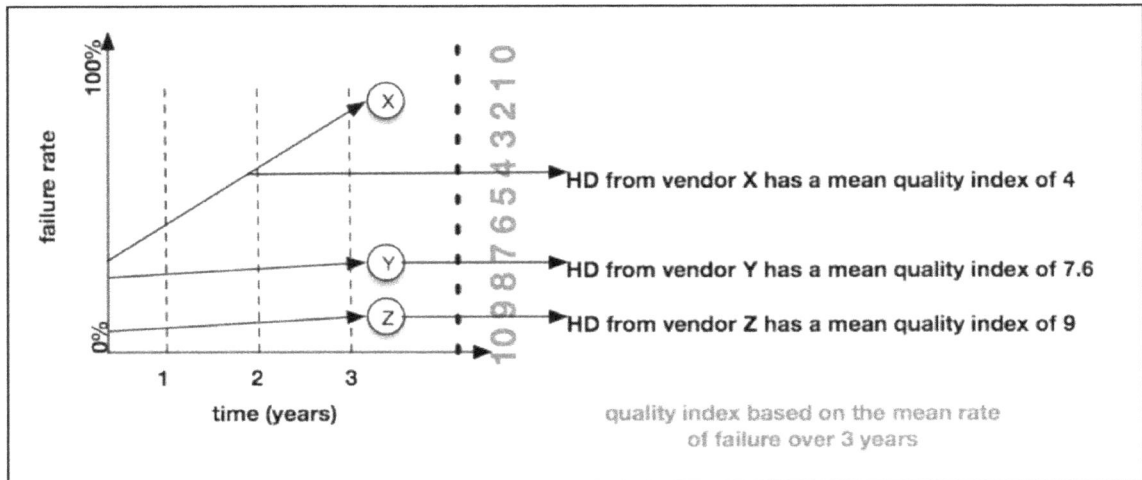

Figure 3-2. Vendor drive quality index

If you were to compare the cost of each of these drives, you would see the following.

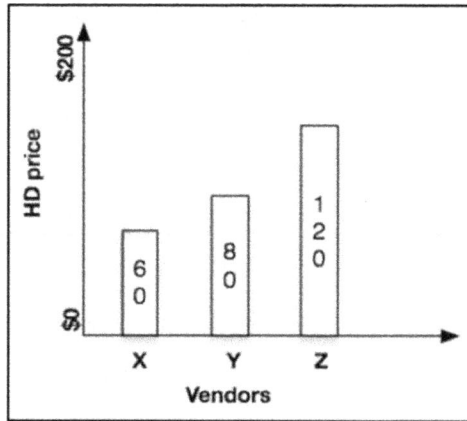

Figure 3-3 Vendor drive cost

Manufacturer	Size	Cost
Vendor X	2TB	$60
Vendor Y	2TB	$80
Vendor Z	2TB	$120

Now let's have a look at a graph that represents how far our money can go and the quality of the drives we get. If we had $1000 to spend, how many hard drives could we get for that amount of money?

For $1000 we could get the following:

8.3 hard drives with a quality index of 9.

12.5 hard drives with a quality index of 7.6.

16.6 hard drives with a quality index of 4.

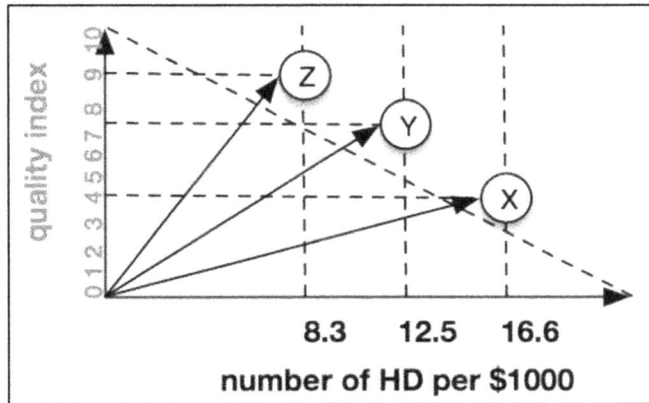

Figure 3-4. Number of drives per $1000

Looking at the details, we could get twice as many hard drives from vendor X than vendor Z, but we know that the failure rate is 6 times as much.

So back to opportunity cost.

The basic formula for opportunity cost is:

The opportunity cost = most lucrative option – chosen option

In this instance if we relate it to the hard drives, we would take the most lucrative option in each drive comparison and see what was given up for the choice. Our sample set has price and QI as the two options being measured.

Opportunity cost of choosing vendor Z over vendor Y is $40 per drive.

Opportunity cost of choosing vendor Z over vendor X is $80 per drive.

Opportunity cost of choosing vendor Y over vendor Z is a QI shift of 1.4.

Opportunity cost of choosing vendor Y over vendor X is $20 per drive.

Opportunity cost of choosing vendor X over vendor Y is a QI shift of 3.6.

Opportunity cost of choosing vendor X over vendor Z is a QI shift of 5.

Now we have to determine if the costs saved by price alone are more beneficial than the sacrifice of quality.

We will assume that the cost of all actions involved with remedying the failure of a hard drive (which is non-business application impacting) is $320, including time from operations, communications, RMA, management, etc.

(This is calculated as 4 hrs operations time (@ $50/hr) and 2 hrs management time (@$60/hr))

Now let's calculate the cost of the failures in a year, based on the number of drives per $1000.

From vendor X we have 16.6 drives, with a failure rate of 60% = 9.96 (or 10, because a partial failure is still a failure). 10 drives at a replacement cost for resources at $320 each, is $3200.

From vendor Y we have 12.5 drives, with a failure rate of 24% = 3. At a replacement cost for resources at $320 each, the cost is $960.

From vendor Z we have 8.3 drives, with a failure rate of 10% = .8 (or 1, because a partial failure is still a failure). 1 drive at a replacement cost for resources at $320 each is, well $320.

If we look at the TCO for the hard drives over a three-year period, it is very telling.

Vendor	Size	drives /$1000	RAW	TCO	TCO/TB
Vendor X	2TB	16.6	33.2TB	$4200	$126.50/TB
Vendor Y	2TB	12.5	25TB	$1960	$78.4/TB
Vendor Z	2TB	8.3	16.6TB	$1320	$79.51/TB

drives/$1000 = number of drives per $1000
RAW = Total raw space
TCO = Total cost of ownership including operations and capital cost
TCO/TB = Total cost of ownership per Terabyte of raw space

TCO per TB for each vendor drive.

The total cost of ownership for the drives is almost the opposite of what is shown when just looking at the capital cost. The reason is because the costs were hidden in operations time. Vendor Y and Vendor Z drives have almost the exact same TCO /TB value, whereas Vendor X is 60% more expensive.

That scenario was for a non-business application impacting failure. It is said that the average business affecting failure costs about $10,000 in lost revenue per incident (which is 31 times the amount for a non-business impacting one).

With a higher rate of failure, the chances that a failure will be business affecting are much greater. Concurrent failures also compound problems because of the higher chance of losing more data at once.

Lastly, we will go back to opportunity cost again, looking at it subjectively.

The opportunity cost for purchasing the cheaper drives is the amount of time that could have been put towards other operations activities and processes. In the end, the solution with the cheapest capital cost has the highest total cost of ownership by far.

So, be careful when looking at the cheapest options, because they are often not what they seem.

3.4 How success can destroy an organization

The definition of success can operationally mean "keeping the lights on" or in accounting terms, "staying in the black". It can also mean consistent growth, year over year. When success is expected, it attracts investment but it also attracts sharks.

When a lot of success happens rapidly in a short period, it can be a due to the perfect succession of events and careful planning, or a windfall from a high risk / high gain decision, or it could be from a market bubble.

Becoming successful and staying successful are two very different things though.

In 2008 the Bank of Korea published a study on companies world-wide, that are over 200 years old. The study found that there where 5,586 companies of that vintage. Interestingly, 56% (or 3,146) of those companies are based in Japan.

In Japan, a large number of companies are family run and passed down through the generations. The oldest company in the world, which passed down through 40 generations (until 2006 when it became a subsidiary of Takamatsu) was Kongo Gumi. It was founded in the year 578 as a construction company that built temples.

Businesses that last the test of time have similar qualities, but with two major delineations:

1. They are lean and lasting.

or

2. They are growing and lasting.

The lean and lasting companies have a very singular focus. They will do one thing and do it very well. A good example of this is Gallet & Co. It founded in 1466 and is the world's oldest manufacturer of luxury watches. The number of staff is only a few hundred people and succession is only within the bloodline.

The growing and lasting companies have a diverse portfolio of offerings. A good example is Hitachi, which, at over a hundred years old, produces TVs, nuclear power plants and backhoe excavators. They will have a corporate succession strategy, but the major stockholders will often be the founding family.

This is not to say that to have a lasting and successful company, you have to have a family owned organization, but two-thirds of fortune 500 companies are family owned.

So, what is different about family run businesses that allows them to stay in the game so long? The qualities they have in common are loyalty, commitment, involvement and collectivism with the organization and to the group (peers). In Japan, these traits are embedded in society and not only desired, but expected.

To better understand this, we will look at a theory which evolved from the 1970s, called "agency theory". In agency theory, there are two parties; the principle and the agent. The agent is able to make decisions on behalf of, or that affect, the principle. If the two parties have different interests, and the information flow is not pervasive, then the agent may act selfishly at the expense of the principle.

An example of this is when a company hires an employee. The company is the principle and it requires the employee (agent) to perform a job function. The job function that they are hired to do is aligned to the business goals of the company. However, if the desires of the agent and principle are in conflict then there is a problem.

Let's look at it from a different angle. A boy and his sister wanted some ice cream (let's call them Prinn and Aggie), but only one can go to the ice cream truck. Prinn likes the flavor Tiger Tail, whilst Aggie likes Double Chocolate (which Prinn hates). Prinn gives Aggie $5 to buy both ice cream cones. She goes to the ice cream truck and sees that they have the following flavors; Tiger Tail for $3, Double Chocolate for $3 and baby sized Vanilla cone for $2. There is also a deal for two Double Chocolate cones for $5.

Aggie does not want the vanilla, but she also doesn't want it to look like she got a bigger cone.

Aggie comes back smiling with two Double Chocolate cones and says "They only had Chocolate. If you don't want yours, I'll have it"

So, in this scenario, Prinn is the principle and Aggie is the agent.

Figure 3-6. The agent indulges in the chocolate deliciousness of deceit

The agent acted in her own self-interest to get the Chocolate cone. The desires of the principle to get the Tiger Tail cone did not align with the desire of the agent to get the Double Chocolate cone. The agent had the ability to make decisions of behalf of the principle. There was no oversight or ability to verify the claim of the agent.

Aggie was happy, but when she wasn't looking, her favorite dolls ended up in the microwave with Prinn smiling mischievously nearby.

If the desires of the principle and agent are not aligned, then there will be a lack of cohesion, trust and loyalty. Retribution is not a staple in the principle-agent relationship, but misalignment of desires can sow the seeds of division and discontent.

Let's look at a real example and how it relates to success.

Nortel was at one time, the world's largest supplier of telecom equipment. They are also one of the best examples of a miserable failure based on the principle-agent issue.

During the late 1990s, Nortel had a period of rapid expansion that was heavily funded by selling stock and accruing debt. Stock speculators drove the value of the stock upwards to extremely high levels, creating a bubble. The actual revenue was much less than what the stock alluded to, which caused the stock to plummet. In October 2000, the stock lost half its value in a single day.

During the decline of Nortel, senior management and execs falsified accounting records which stated that the company earned hundreds of millions more than it did. They then proceeded to pay out massive multi-million dollar bonuses based on these falsified records.

The desire of the principle (stockholders wanting a consistent and improving return) was not aligned with the desires of the agents (execs wanting personal bonuses). There was not sufficient oversight on the part of execs to ensure that they were acting in good faith for the stockholders.

There was also another diversion between the principles and the agents, and that was risk appetite. The process of rapid expansion required a higher risk appetite than was experienced over the course of the previous century that Nortel had been around. The successes changed the priorities of the execs and the overall direction and dynamic of the company. This ultimately led to the situation that was their downfall.

Small companies that are successful are often able to thrive because of key people that wear many hats. They have a wide array of talents and are able to jump into many roles, making them more flexible and quicker than larger companies. The CEO could also be marketing, accounting and the janitor whilst the COO could be the helpdesk, business analyst, operations department and lead coffee machine programmer.

The resources of individuals are finite, thus a growth in business will require more staff and a greater division of responsibility. As the company changes, the principles and management have to evolve as well, giving the reigns to their role successors. The dev-ops team can handle the coffee maker so that the COO can focus on general operations.

If the company's principles and management do not evolve, then you end up with micro-managers. Micro-managers break three important principles of business:

1. Opportunity cost

When lower level jobs are being performed or scrutinized, that takes time away from doing more valuable work that could be advancing the business.

2. Comparative advantage

When lower level jobs are being performed or scrutinized, the efficacy and performance of the role that they are responsible for is diminished.

3. Authority is tied to responsibility

If managers hold subordinates responsible for doing a job, then they should also be able to accomplish it as they see fit.

These three things can destroy employee morale, dampen initiative and hinder the ability for employees to move up the corporate ladder. A company with micro-managers will find that roles will more often than not, be filled by external candidates in an ever-cycling manner.

The last thing we will discuss is what is referred to as a "culture of arrogance". When a company is a market leader, or a big fish in their pond, they will continue doing what they are doing because it is working for them. They will lack innovation and become complacent. There was a saying in the industry," nobody ever got fired for buying IBM". This is a perfect example of the kind of complacency that will ruin a business.

IBM has been on a steady decline because of its lack of innovation and ability to adapt to market changes. This was very apparent in 2013 when the CIA put out an RFP for its cloud computing business. Historically, this is the kind of domestic contract that IBM would always win. The contract was worth 600 million dollars and IBM was bidding at 30% lower than the competitor, which was Amazon Web Services. The CIA went with Amazon because IBMs solution was technically inferior. The CIA stated in its justification, that it was "buying innovation".

If a company rests on its laurels, it will find that opportunities will pass them by and they will then be in a decline towards obscurity.

So, to recap, success of a company can cause the following:

1. Divergence of interests between the principles and the agents.
2. Risk of spreading resources thin and creating a feeling of discontent from micro-managing.
3. Creating a "culture of arrogance" and slowing innovation, which allows competitors to overtake market share.

3.5 Organizational Survivability in the aftermath of an incident

When a company has a major failure, how long can they survive offline before the company is unable to recover?

USA Today surveyed 200 data center managers and found that more than 80% reported that downtime costs exceeded $50,000 per hour. For more than 25% of the managers, downtime cost had exceeded $500,000 per hour.

In August of 2013, google had a 5-minute outage on all their services. It caused a 40% drop in all internet traffic globally and cost the company $470,000. That's nearly 10 Million dollars an hour.

When a company is conducting a business impact analysis (BIA), it will determine the RTO/RPO (recovery time objective / recovery point object) of all systems within the infrastructure. One item of note is the MTPoD, which is the Maximum Tolerable Period of Disruption.

The MTPoD is the duration of downtime that will cause a company to no longer be viable. If a company is down for long enough, it may not be able to come back online.

Major outages may also cause a domino effect that can put a company out of business within months. Similar to a life-threatening injury, the company may bleed out on the operating table.

Some things to consider when an outage occurs:

- What is damage to the company's reputation?
- In the case of data corruption, being back online does not mean back in business.
- Public perception and loss of confidence will decrease the stock value.
- There will be losses of business opportunities.
- There will be a loss of employee morale.

Depending on how a company is structured organizationally will also determine the degree that they are tolerant to critical outages.

Some questions to address when examining the outage tolerance, are listed below:

- Are all IT operations centralized, or are they distributed?
- Can a remote site operate independently if the head office is down?

- What happens in the event of a communications failure?
- How can a problem be resolved if there is no communication between locations, departments or staff?
- Are there alternate channels of communication?
- What is the process for the alternate communication?
- Has the process been socialized?

I have seen a company that had an in-house mail server, web server and phone system. They had a power outage and all three went down. Mail started bouncing after a few hours and they couldn't call anyone to fix it or give an update to management. Their corporate website was offline and local cellular services were overloaded so they couldn't make any outbound calls.

Land lines were still working, but no POTS (plain old telephone system) phones where not anywhere to be seen. In the end, they were able to get an old laptop with an internal modem and use a dialup internet connection and a personal email account to communicate with management.

The outage which lasted a couple of days, severely affected client confidence and they lost 50% of their business over the next few months, which forced layoffs and other cost cutting measures. In 2 years, they were acquired by a competitor for 1/10 of their original value.

3.6 Risk accountability and transparency

Who is ultimately responsible for the risk within an organization? Is it operations? Senior management? Each individual staff member? Is it the last person that touched something?

If a network administrator logs into a switch and changes a port policy, and then shortly thereafter all IP voice communications stop, is it their fault? Not necessarily, but more often than not they are guilty until proven innocent.

That is an issue of correlation versus causation. What if they are proven guilty of the configuration change that cost the company tens of thousands of dollars in downtime?

Was it the fault of the network admin for making the change? Their direct actions caused the outage.

What about if there was a change order that the admin was just following? Then this moves the accountability away from the admin, because they were just doing their job.

What if that change order was not put through the required change management process? Does this mean that the admin is now responsible? Or was it the requester?

What if the risk of voice communications going down because of a port policy change was known by the unified communications (UC) team, and it was on their "get around to" pile to fix and socialize this information? Is it the fault of the UC team for not sharing this risk information?

What if the UC team setup the voice system incorrectly which caused ongoing network disruptions?

What if the CEO told the network administrator to fix the network problem right away and bypass change management because the disruptions are causing lost revenue? The network admin easily found the problem. It was a simple port policy change!

The importance of whose fault it is starts to diminish after a few rounds of the blame game. Looking for a scapegoat is not a valid response for a failure.

What is important is who should be vigilant to the day-to-day risks that they are directly associated with. Once a risk has been identified, the people that do the day to day management of those resources are called the *risk owners*. They will monitor any changes that occurs to that risk. If an action has been created to address the risk, then it is performed by the *treatment owners*. The action that will be taken is decided by the *control owners*.

The accountability for the risk goes to the risk owners.

The treatment that is to be performed is decided by the control owner.

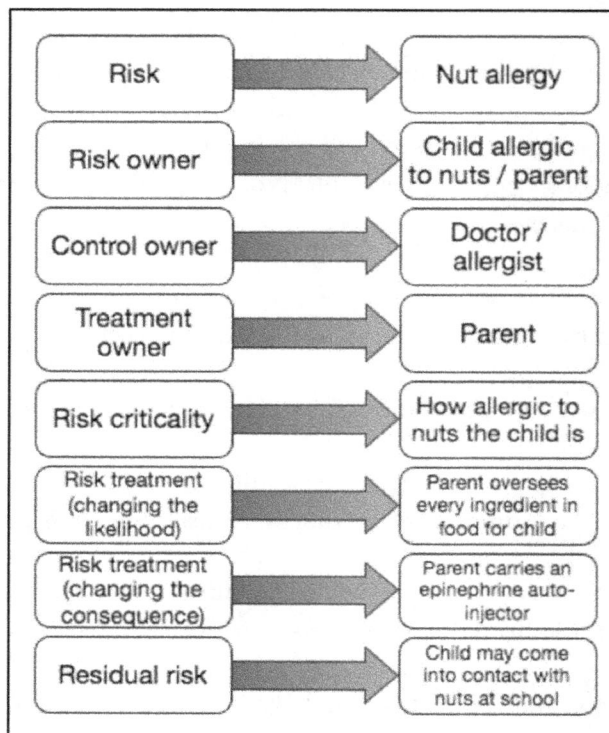

Figure 3-7. The risk treatment response and culpability

The action for addressing the risk (performing the treatment) goes to the treatment owner.

If a risk has been identified and a treatment course has been created for it and it has been implemented effectively, then the chances of that risk occurring have been greatly reduced. However, there is always a residual risk unless the action, process, or technology that is at risk is removed entirely.

I'll use the example of a nut allergy for illustration purposes.

If a child has a nut allergy, that is an identified risk. How severe the allergy will determine the criticality of the risk. If they contact nuts, will they get a stuffy nose, or will they go into anaphylactic shock?

The child with the nut allergy is ultimately the risk owner, however depending on the age of the child, the parent may also be classified as the risk owner.

The doctor that has recommended a treatment is the control owner.

As a preemptive measure, the parent may examine the ingredients of every food item they buy and they may also carry an epinephrine auto-injector (epi-pen) as a reactive measure.

The parent is initially the treatment owner, however as the child grows up, they will become the treatment owner.

Below are different types of risk treatments that can be applied in the nut allergy scenario:

1. Avoiding the risk.

Do not go to places that may have nuts. Do not contact people that consume nuts. Do not buy foods that may have been processed in places that have been in contact with nuts.

2. Taking the risk to pursue an opportunity.

Going to school increases the likelihood of encountering nuts, but it is required to get an education and socialize with peers.

3. Remove the risk source.

Remove all contact with the outside world.

4. Changing the likelihood of the risk.

Limit possible contact with nut exposed areas. Ensure that the school has a nut-free policy, is socialized and is enforced.

5. Changing the consequence.

Carry an epi-pen to counteract anaphylaxis.

6. Sharing the risk with another party.

The parent shares the risk ownership with the child.

Retaining the risk by informed decision.

If the criticality of the risk is low enough, then understanding the effects and responding to them ad-hoc may be done. The child may not change what they eat or how they interact with the world, but if they get a reaction then they will re-evaluate what they are doing or eating.

In order for risk to be addressed in the most efficient manner, it needs to be identified and then communicated to the risk owners. Once the risk owners have this information, it then needs to be communicated to all stakeholders via a defined process. The stakeholders or risk advisory board will then deliberate and the selected or predefined control owner will provide the treatment.

This all sounds fairly straightforward in theory, but how does it look in practice?

Looking at industries that have volatile markets, interconnected global economies, competitive pressures and regulatory requirements, this becomes a lot less black and white.

If a risk is identified, but the socialization of that risk will cost more than the effect of the possible risk (if it occurs), then should it be disseminated?

If a risk has been identified, but it's likelihood is currently extremely low, should it be a topic of conversation?

If a risk has been identified by a third party, what obligation do they have to communicate it with the risk owner?

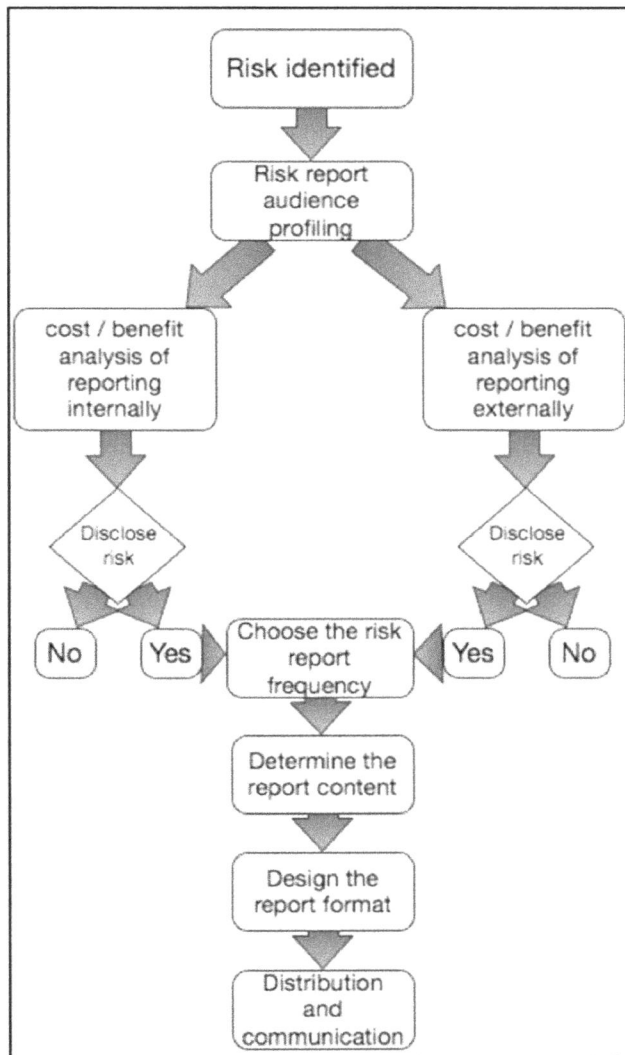

Figure 3-8 – The risk reporting process.

These are all questions that have moral implications and unless there is a prescribed way to address them, the people making these decisions will often factor in how the decision will affect them. Standards in transparency, no-fault whistleblowing and companywide open-door policies with management help to ease the decision process.

The truth in practice is that individuals that identify risk are not usually the first ones to learn about them within an organization. The risk may already be known, but a different treatment action may have been decided on. If the treatment action that management has decided on, differs greatly from what the individual supported, then there may be an unresolvable conflict.

If the decision of management negatively impacts public health, safety or wellbeing, then the individual can voluntarily report the risk to an appropriate external third-party agency. This will make them a "whistleblower", and there are laws to protect the individual from retribution. In a study that looked at whistleblower cases from 1994 to 2009, 74% of whistle-blowers were found to have been fired, 15% received poor performance evaluations, 6% received suspensions and 5% had transfers. Only 2% were reinstated in their old jobs and 8% won damages.

Internal resolution should always be the first route for action and reporting. External reporting can also put the organization at further risk by publicizing risks that have not do not have a treatment yet.

Every scenario is different, but the reporting process can be standardized. The diagram above shows the workflow process for decentralized risk reporting that can be used by every level of an organization.

For an organization to be considered transparent, the recipients of risk information will have:

- An accurate representation of the risk and its underlying components
- An understanding of how the current risk profile compares to previous periods
- An explanation for what has accounted for any changes that have occurred

3.7 LD50 of Risk appetite

This is not an actual term, but I will convey my reasoning for expressing it as one. My definition for an LD50 of risk appetite is "when the amount of risk accepted by an organization has a 50% chance of leading to its unrecoverable decline in the immediate future".

In medical terminology, when a toxin is introduced to a population and the concentration is sufficient to kill 50% of those affected, that is the LD50 (LD standing for lethal dose). A specific amount exists for just about every substance out there. There is a saying in toxicology: "the dose makes the poison". To reach an LD50, it would take; 175 espressos or 13 shots (45mL each) of Vodka or 6L of water. Yes, plain old water can kill you by causing Hyponatremia, which is a low concentration of sodium in the blood. This is what killed famed artist Andy Warhol.

Figure 3-9 If Julius Caesar was a modern CEO.

There can always be too much of a good thing.

When taking risks, there is always a point at which the potential benefits are overshadowed by the potential losses. This can be on a single decision or on a series of decisions.

In gambling, when a person reaches what is called the "point of no return", they have lost enough money that the chances of recouping any losses have faded to obscurity and they must deal with the reality they have created. A series of decisions has led them to where they ended up, each individual decision was not necessarily risky in and of itself, but together they created a state that had the potential for great rewards but instead they experienced great losses.

In 49 B.C., Julius Caesar crossed the Rubicon river with a legion of his army. This was a boundary that was forbidden by law to be crossed by a general leading his army, as it would be seen as an act of insurrection, punishable by death. This single act has survived in the idiom "crossing the Rubicon", which means that there is no turning back. The Rubicon is the point of no return.

A full commitment is required for success; the alternative being dire consequences not just for the individual, but for all people involved.

In both the gambling example and the one with Julius Caesar, there was a driving passion that put aside logic and reason for a vision. Though one is desperation and the other is entitled inspiration, they are more similar than different. Both examples achieved the LD50 of risk appetite.

In business, leadership of an organization is not exactly the same as leadership of a Roman army because the threat to human life does not exist. However, the threat to the life of the organization is in the balance. The same is true for the oversight of IT; key decisions can make or break the organization. The alignment of risk appetite between business units is fundamental to ensure that the maximum benefit is gained from the risks taken.

If the risk appetite for Caesar and his army were not aligned and some of his troops said, "You go ahead. Let us know how it goes," or if they ran ahead in a berserker rage, then their risk appetite would not be aligned with management and history would have been different.

It is important to understand motivations and why an organization is being more or less tolerant with the risks it is taking. If these motivations are not shared across the business units, then understanding the vision can be difficult and risk alignment will not be there.

Sometimes the risks that are taken are not fully understood by the organization leadership. There may be a few key advisors from the different business units that bring this information to the table. It is not uncommon for each business unit to maintain its own risk register and then disseminate that to the leadership group on a regular basis.

If the leadership does not have a full comprehension of the scope of a risk, it makes it impossible to make decisions that address all variables. This means that there is a chance that a lack of understanding or available information may put the organization in a situation where an LD50 of risk appetite exists.

The leadership of an organization is also a human process, which means that there is a finite amount of attention that can be given to risks. If the amount of attention that some risks are getting are more than their share of the total attention available by the leadership, then there may be things that are missed. Risks that are quite severe may be hidden because the multitude of less severe risks get the spotlight.

The response of leadership to identified risks will also determine how often the they are informed by the business unit advisors. If the party that identifies a risk is automatically responsible for the treatment of it in addition to their regular work load, then they are less likely to volunteer that information.

3.8 Methods for determining probability

When determining probability, you can do so with the precision of a shotgun or a scalpel. With a shotgun, you point in the general direction you want and you'll probably hit what's in that area. With a scalpel, you can make an incision that will probably be between 0.6mm and 1mm deep.

If you were to use the shotgun method in determining risk probability (also known as subjective Bayesian probability), then your probability would be a scale of 1-5 (1, which is not likely to 5, which is definitely). The process to determine the value would be how you *feel* about it.

An example would be: The probability that the cable repair person arrives on time is 2 out of 5. The reasoning for that is that they have never been on time, ever, but they get a slight benefit of the doubt, which brings them up to 2 out of 5. Since there is some history that comes into the decision-making process, it can be considered a "Calibrated probability assessment".

If you used the scalpel method in determining risk probability, then you deal with information and numeric values.

John Maynard Keynes wrote a book in 1921 called "A Treatise on Probability" which is considered one of the most important works written in the field. He surmised that there were three domains of probability:

1. Frequency Probability
2. Bayesian Probability
3. And "special cases"

Frequency Probability is defined by an event that has only two possibilities; it happens or it does not happen. The number of trials in which the event has a chance to happen is called the sample space (identified by nt). The number of times in the sample space that an event happens is called the occurrence (identified by nx). The probability of the event happening (identified by P(x)) is approximated to the number of occurrences divided by the number of trials in the sample space.

$$P(x) = \frac{nx}{nt}$$

The more trials that are sampled, the higher the accuracy of the probability of the even occurring. An example would be:

The cable repair person was supposed to come five different times, but only showed up once.

P(x) = The probability that the cable repair person will show up (which is 1/5, or 20%).

nx = The number of times that the cable repair person showed up (which was 1).

nt = The number of times that the cable repair person showed up (which was 5).

There is a 20% chance that the cable repair person will show up based on frequency probability.

We talked about *subjective* Bayesian probability, now let's talk about *objective* Bayesian probability. Bayesian probability measures a degree of belief, and thus must begin with a hypothesis (identified as H). The hypothesis is then supported or countered by evidence (identified as E). Probability is identified as P.

$$P(H \mid E) = \frac{P(E \mid H) * P(H)}{P(E)}$$

This method of determining probability allows for the additional information to be added, thereby changing the prediction.

| represents a conditional probability, it means "given".

P(H), is the prior probability, is the probability of the hypothesis (H) before the evidence (E) is observed.

The evidence (E) represents new data that was not used in determining the prior probability.

P(H | E), is the posterior probability, is the probability of the hypothesis (H) given the evidence (E).

P(E | H) is the probability of observing the evidence (E) given the hypothesis (H). As a function of the hypothesis (H) with the evidence (E) fixed, this will be the likelihood.

P(E) is considered to be the normalizing constant. This factor is the same for all possible hypotheses being considered.

The posterior probability (P(H | E)) of a hypothesis is determined by a combination of the inherent likeliness of a hypothesis (the prior probability), which is P(H)) and the likelihood.

I will not outline an example with the math for this, but rather the intent of the equation, as I believe it better reflects the goal of the method.

An example would be:

We don't think the cable person will show up because of past experiences.

This is the prior probability P(H).

Below are listed the known facts that we can apply to the scenario. This is the evidence (E):

Facts that weigh towards the cable person not showing up:

- Historically they have not shown up on the first few appointments.
- The cable company overbooks appointments because of the possibility of cancellations.
- Appointments often go over the allotted time.
- The cable persons first appointment of the day and immediately after lunch time are the most likely appointments to occur because overbooking requires that some appointments are missed.

Facts that weigh towards the cable person showing up:

- The chances that the appointment will be met increases with the number of failed appointment in the past. If this wasn't the case the cable company would go out of business.
- The location is in a centrally populated area that is not very busy during the day and is easy to navigate to.
- Complaints to customer service may have increased the priority in the service queue for the appointment

Each one of these facts or "items of evidence" have a likelihood that they indicate the occurrence or lack of occurrence of the event.

This is the P(E|H), or likelihood of a direct correlation of the evidence to the event.

> Thank you to the local cable company for providing me with a scenario for describing probability. Without finally showing up, the example would not have worked as well (and neither would have my internet).

There may also be a case where the evidence is so compelling, that it supersedes all counter evidence. This is called a *fortiori* argument. An example is this:

There is an earthquake and all power in the region is cut and roads are blocked. This evidence is a very strong argument that the cable guy will not show up on time, and it supersedes the determination of the company for good customer service.

A "special case" in probability is when:

- You have limited or no prior instances of an event, so frequency probability cannot be used.
- There are many possible interactions that will significantly affect the possibility of the event.
- There are many possible outcomes for a situation.

For cases such as this, you need to map out all the possible outcomes and then determine all the interactions that may sway the chances of a possibility. The possibility of a specific outcome happening can be approximated using a complex probability theory, which unfortunately is out of the scope of this book. Mainly because I cannot explain how to use it effectively in simplistic terms.

How a probability is recorded depends of what it is used for. A qualitative value can be used for most scenarios in order to provide a human readable report which can be used to instigate actions. A quantitative value can be used for further analysis. A quantitative value can be changed to a qualitative one, but doing the inverse would lead to great inaccuracies.

Below is a qualitative probability scale.

Simple probability scale

Rating	Descriptions
A	Extremely unlikely
B	Remote possibility
C	Occasional occurs
D	Reasonably Possible (repeated failures)
E	Frequent (failure is almost inevitable)

Below is a more precise quantitative scale:

Detailed probability scale

Risk Event Probability	Interpretation	Rating
> 0 - <= 0.05	Extremely sure not to occur	Low
> 0.05 - <= 0.15	Almost sure not to occur	Low
> 0.15 - <= 0.25	Not likely to occur	Low
> 0.25 - <= 0.35	Not very likely to occur	Low
> 0.35 - <= 0.45	Somewhat less than an even chance	Medium
> 0.45 - <= 0.55	An even chance to occur	Medium
> 0.55 - <= 0.65	Somewhat greater than an even chance	Medium
> 0.65 - <= 0.75	Likely to occur	High
> 0.75 - <= 0.85	Very likely to occur	High
> 0.85 - <= 0.95	Almost sure to occur	High
> 0.95 - < 1	Extremely sure to occur	High

An important thing to consider is the philosophical question; if a tree falls in the forest and no one hears it, does it make a sound?

Or in this case; if an event happens and no one detects it, how do you know it happened? The probability of detection should also be determined for an event. A scale for this is as follows:

Rating	Meaning
1	Certain - fault will be caught on test
2	Almost certain
3	High
4	Moderate
5	Low
6	Fault is undetected by Operators or Maintainers

The time between when an event occurs and when it is detected is called the dormancy period. The dormancy period could be seconds, days or months, depending on how the event manifests. If an event occurs and it immediately impacts users, then you will find out about it with high probability in a very short dormancy period. If it doesn't affect users immediately, it may be putting the environment in a risky state that could compound any future events into a critical state.

The detection probability can be determined by understanding monitoring and maintenance cycle in your environment. Over a period, every component will be analyzed and polled at some level. If this is not the case, then that is a large risk in itself.

3.9 Risk vectors

Risk vectors are the means in which risk can manifest as an event in an environment. A simile to this would be that mosquitos are a disease vector. They are the method that a disease can be transferred by, manifesting a disease in a biological environment.

Humans are a risk vector.

As soon as humans enter the equation, it introduces an element of entropy to a system. There are methods of determining approximate entropy a person can have on a system, but I will get into that in a later chapter.

Strict processes or lack of process is a risk vector.

By having a defined method of doing things that does not account for every possible scenario and cannot adapt, the response cannot be calculated and thus it is possible for non-ideal response mechanisms to kick in. An example of that is when someone jumps out of a closet to scare a friend, and then they get punched in the face. The fallback defense mechanism was non-ideal.

A lack of process creates an ad-hoc response method that is difficult to replicate. Doing things ad-hoc is not efficient, scalable, nor does it provide quality results. Try doing something spontaneous, several times in a row. It is very difficult to accomplish because process has a tendency to creep in, so it's better to just welcome it. If a process is created, then there is a greater control of cause and effect.

If an organization has consistency in its deliverables, then it can conduct measurement of performance and strategize ways to improve. If there is no consistency, this is usually due to lack of process. Many small organizations and startups experience this and it will inhibit the chance to grow the organization or maintain a loyal customer base.

Technology is the most common risk vector.

This can comprise the design, the components, reliability, dependencies, etc. Risk vectors can be organized at a high level into people, process and technology. However, each area can be broken down into much more specific sub-groups. Below are those subgroups, within each risk vector. There are 13 sub-groups:

People

1.   Budget Risk: If the project is not scoped properly, then unaccounted delays of delivery may occur. This may even cause the entire project to flop.

2.   Operational Risk: This can occur if there are insufficient technical resources, or if they have insufficient training. It can also occur if there are multiple concurrent activities that need to be performed and critical ones are not prioritized.

3.   Resource Risk: If staff resources are not scheduled properly, or if budget is not managed effectively then this creates resource risks.

4.   Supplier Risk: When an unmanaged third party has a role within a project, a lack of delivery oversight can create risk in quality or in being able to meet timelines

Process

1.   Infrastructure Risk. Poor planning and implementation may lead to sub-optimal installations and intermittent performance.

2.   Programmatic Risks: This is when risks come into play because of deviations past known or recommended thresholds.

3.   Technical Environment Risk: This is when the work environment is in a state of flux and certain expectations or processes are not adhered to.

4.   Business Risk: If tasks such as business document workflow are not followed in the prescribed order, then delays and miscommunications will occur.

5.   Quality and Process Risk: This can occur due to lack of training or direction for employees, or by lacking performance metrics to know when process deviations occur.

6.   Schedule Risk: If there is no effective project management office (PMO), then deadlines and priorities have a risk of being missed. This can lead to cascading impacts on other projects or initiatives, financial impact and ultimately project failure.

> ### Technology
>
> 1.   Information Security Risk: If intellectual property or confidential data is leaked then there is a risk of liability to the service provider / operations unit, and a personal or business risk to the data owner.
>
> 2.   Technology Risk: This can occur if technology is being implemented that is unproven, under-tested or is being deployed in a use case that it has not been validated for.
>
> 3.   Architectural Risk: This risk category occurs when there is a performance or functionality impact due to a flaw in the design.

3.10 Fault Tree Analysis

Fault Tree Analysis (FTA) was devised in 1962 at Bell Laboratories as a means to evaluate the control systems of the "Minuteman" inter-continental ballistic missiles (ICBMs).

It is a top down, deductive failure analysis method that uses Boolean logic to combine a series of lower level events into larger events. It is useful as a tool to analyze possible cascading effects of failures and determine the best areas to put resources in place to bolster, isolate or modify systems or components.

FTA is frequently used in nuclear power engineering, aerospace, chemical processing and the pharmaceutical industry.

The top event is referred to as the "system failure condition". The severity of the system failure condition will determine the depth of the FTA.

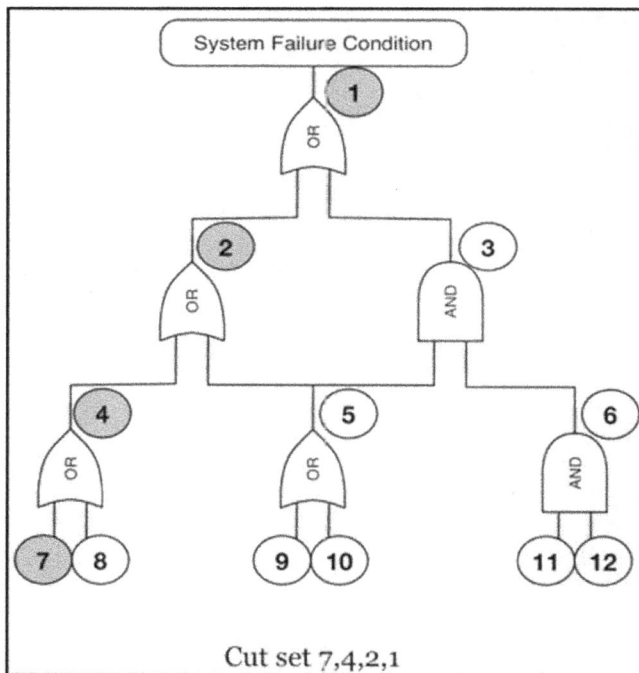

Figure 3-10. An FTA system failure

FTA can also be used for the following:

- To understanding the event logic that could lead up to a system failure condition
- To prioritize risk responses on the main contributors leading up to a system failure condition
- To understand the current system state and critical dependencies during a failure event
- To minimize and optimize resources
- As a design tool, it can be used to create functional requirements
- As a tool to aid with a Failure Mode Effect Analysis (FMEA)

Boolean logic is comprised of what are known as "gates" that determine the end state, or output from inputs. It is the same way that digital circuitry works, with 1's and 0's. 1 is equal to an "on" state and 0 is equal to an "off" state.

The logic gates that are used in FTA are called AND & OR.

With an AND gate, all inputs need to be 1 for the output to be 1.

With an OR gate, any of the inputs could be 1 for the output to be 1.

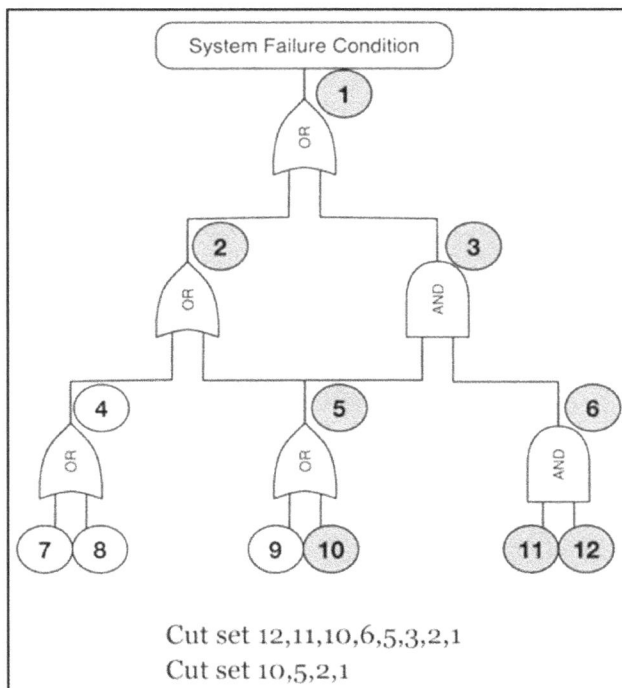

Figure 3-11. An alternate FTA System failure

In relation to an FTA, the 1 signifies an event happening and 0 signifies that it does not happen. When a series of smaller events occur or don't occur, the state of the system can be mapped out with an FTA. This will show you what needs to be fixed immediately at each failure state and what the next probable failure will be.

Sometimes multiple areas are affected by a single event. In this case, the event will show up multiple times within the fault tree. This is what is referred to as a common cause.

The path through a fault tree to the system failure condition is called the "cut set". The shortest possible path through a fault tree is called the "minimal cut set".

A fault tree starts from a system failure condition and flows down to all the initiating events. An alternative model, called an event tree can also be used. This starts with an initiating event and branches out to all possible outcomes. It is a bottom up approach that can be quite complex and have many branches that reach many different system failure states.

Some industries use both models, whereas others use just one.

The complexity of the analysis models requires special software to map out the branches, cut sets, probabilities of those cut sets and of the individual initiating events. In the US, many nuclear plants and aerospace companies use software from the Electric Power Research Institute (EPRI) called CAFTA. Internationally, the software known as RiskSpectrum is used by almost half the world's nuclear plants.

CAFTA stands for Computer Aided Fault Tree Analysis System. IT has been used for over 30 years by the nuclear and aerospace industries to decompose complex systems into subsystems and components that can be evaluated against known root causes. This allows for the visualization of component interaction and dependencies during many states of operation and levels of failure.

3.11 Determining Acceptable Risk

This concept relies on many human factors such as leadership direction, personality types, opportunities and risk appetite. It is impossible for everyone to have the same idea of what is acceptable risk.

Certain criteria can be used to make a standard determination whether a risk is acceptable or at a minimum, tolerable.

A risk can be acceptable if it meets the following criteria. Not all are required, but they can be used as general guidelines:

1. It is below an already accepted threshold.
2. The probability is below an accepted threshold.
3. It would be cost more to reduce the risk than the cost of the risk itself.
4. Multiple subject matter experts have validated that it is acceptable.
5. Popular industry opinion deems it as acceptable and there is validation for why.
6. The opportunity cost of resources would not be more effectively applied in other areas.

The big boss says it's acceptable and personnel safety is not at risk.

3.12 Black Swan events

The term "Black Swan" originally came from the idea of something that has never happened and is impossible.

However, the meaning had to change when a type swan that was black was actually discovered. The new meaning then was; something that was very unlikely to occur.

Figure 3-12 A black swan.

The "black swan theory" was later created by a mathematician named Nassim Nicholas Taleb. It was used as a metaphor for events that are extremely rare, hard to predict, come as a complete surprise and have a major impact. These events are far beyond the normal scope of consideration for the people affected by it. Examples of this are:

The arrival of Europeans to North and South America: The indigenous people did not expect the arrival of foreigners nor did they understand the concept of land "ownership". It was unfathomable that people could try to own something that was deemed a shared right, like owning sunlight or air.

The French Revolution: The nobles and aristocrats did not see the signs. It came as a total shock to them when the revolutionaries came for them and beheaded them with guillotines. This started a global departure from monarchies leading countries to having political parties and democracies.

The dawn of the internet: In the early 1990s, only schools, scientists and hobbyists were on the internet. A decade later, it had become an essential component to every manner of daily life. Nowadays it it almost inconceivable to not be perpetually connected to the net in one form or another. The impact and scope of this connectivity has dramatically altered the modern world.

The American housing market bubble in the 2000's: Millions of people lost their homes because of a market correction that devalued houses to a fraction of their former worth and the economy plunged into recession as financial institutions collapsed.

In each of these examples, the people affected had no inclination of the criticality, scope or wide-ranging impact that would occur from these events. However, looking back at all the associated events, the signs and the patterns, the causes and effects can be completely understood.

It can be said that people have biases based on the way they think, or see the world that blind them to what may be happening right in front of them.

Black swan events can be summarized by three key points:

1. It is something that is not normally expected under regular circumstances.
2. It has an extreme impact on the people or things it affects.
3. It can be predicted retrospectively by analyzing associated patterns and events leading up to the black swan event.

Black swan events are outliers, that should be considered and planned for. They can be utilized for gain, or you can protect yourself and organization from failure. However, this can only be done by seeing the greater picture.

What is happening right now in the community, the industry and in the world? What are the signs, the patterns, the indicators? What is the next disruption and how will that affect the current state of operations? Is there a change in culture? Politics? Technology?

There is an expression that says, "See the forest through the trees". It means, to see the bigger picture you have to widen your focus. Do not get lost in the minutia of details, or the tasks at hand.

I suggest that the best way to do this is to reflect. Put time aside for introspection and reflection of everything. Draw upon history within the industry, within the organization and personally. Find the patterns, associations and indicators that help to unravel the puzzle of the next "surprise event".

In reality, there are no surprises, only overlooked possibilities.

Taleb not only predicted the financial crisis of the late 2000's, but also profited off it. He also held the position of Distinguished Professor of Risk Engineering at New York Universities Polytechnic Institute.

3.13 Chapter Summary

1. The true cost of high availability needs to be calculated and justified. Not all situations call for five nines of availability. If a situation does require it, then the technical and financial resources need to be in place to ensure it is met.

2. When failures occur, there are many events that happen, each with their own cost. If every foreseeable event that can happen in the aftermath of a failure and their associated costs are calculated, then this will provide more insight into which areas need to be bolstered or modified and what is the best use of money.

3. Knowing the quality of an item or service, based on real metrics, can level the playing field when comparing with others. More expensive is not always better and cheaper isn't always cheaper when the TCO is investigated.

4. Companies that have an inclusive and well defined corporate strategy, will often do better than those that don't. Openness, loyalty, commitment and involvement are key areas that are nurtured in companies that stand the test of time. Being able to identify divergence with corporate interests and personal interests, will allow the identification of that as a risk so a response can be created.

5. A business continuity plan should account for all foreseeable events that may occur and it should have a well thought out response. A risk management strategy should be able to minimize the effects of any identified failures before the business continuity plan engages.

6. The responsibility for risk is spread across many levels of the organization, however everyone takes part in it. Once a risk is identified and a risk response is created, then the level of risk is explicitly set. The response effort and cost will determine the final outcome (or residual risk).

Sometimes the amount of risk that an organization takes on is beyond return. The risk will either become a capitalized opportunity, or it will be a death sentence for the company. It is important to know when an opportunity turns into something prickly, the potential gains, losses and how each stakeholder will respond good or bad.

7. There are several ways of determining probability; frequency probability, Bayesian probability and a subjective "gut" feeling. There is also the probability that something will or won't happen and the probability that it will be detected.

8. The means in which a risk manifests is called the risk vector. There are three main areas; people, process and technology. Below, there are thirteen sub-categories that fall within the main three:

People	Process	Technology
Budget Risk	Infrastructure Risk	Programmatic Risk
Operational Risk	Tech Environment Risk	Info Security Risk
Resource Risk	Business Risk	Technology Risk
Supplier Risk	Quality and Process Risk	Architectural Risk
	Schedule Risk	

9. There is rarely a failure event that is not preceded by other smaller compounding events. It is important to model scenarios to determine the effects of "what ifs". The deeper the layers and mapped out theoretical event trees, the more they resemble the real world. Fault tree analysis and event tree analysis provide a top down and bottom up approach that work together to provide greater insight into the many ways that failures can occur. By utilizing them, one can plan for the entire spectrum of failure states and create risk responses for each.

10. Acceptable risk is something that needs to be defined at an organizational level and acknowledged at all levels. There are guidelines that can be used to create the risk tolerance strategy and communicate it within the organization.

11. There will always be events that are rare, have a high impact and can be rationalized and explained after the fact. The only way to protect against these and even use them to your advantage, is to have the openness to possibilities and see the patterns that predict them.

3.14 Chapter Review Questions

1. A service provider has secured the business of a regional emergency medical services organization. The requirement from the EMS organization is five nines of availability, as it will directly impact the ability of first responders to communicate with dispatch.

2. List the preventative measures that can be performed by the service provider to ensure that the service requirements are met.

3. What will happen if the service requirements are not met and there is an average of 20 minutes unplanned downtime per month?

4. Why is the alignment of risk appetite important within an organization?

5. How does the involvement of management affect the total cost of repairing a failure in an environment?

6. What is the difference between an RCA and an FEMA?

7. You are ordering 100 new servers for a company. In the past, you have worked with a single server of the same model and there were issues with it. How would you determine the probability that there will be issue with the new servers?

8. How would you respond to the risk that there may be an issue with one or more servers?

CHAPTER 4

MacGyver's and Gamblers

"If a magician makes a mistake, it's sometimes forgiven by the audience. If a gambling cheat makes a mistake, they will almost certainly lose their lives - and probably in a horrible manner."
--Steve Truglia (Stunt Coordinator / Magician)

A company's success is determined by its capabilities, its leadership, its culture and the market landscape. Personality types and inter-personal communication determine much of this. This chapter discusses personality types, culture, innovation mindsets and their relation to risk.

4.1 Personality Profile and risk appetite

This chapter is named after the 1980s-television show called MacGyver (with actor Richard Dean Anderson) where the main character could build anything and solve any problem with a bit of ingenuity and a Swiss Army knife. It is in the spirit of that personality type and of people that approach situations from different angles that we base this chapter on.

Personalities have been profiled for as long as psychology has been around. At around 400BC, Hippocrates popularized the temperament theory, which stated that there were four fundamental personality types; sanguine (optimistic and social), choleric (short-tempered or irritable), melancholic (analytical and quiet), and phlegmatic (relaxed and peaceful). This was used for nearly two thousand years until other analytical methods began getting the spotlight.

In the beginning of the 20th century, a well-educated woman by the name of Katherine Cook married Lyman James Briggs, a physicist and the Director of Bureau of Standards. They had a daughter named Isabel whom Katherine home schooled. Katherine was fascinated with knowledge, education and the understanding of personality. She sought to provide the best opportunities for her daughter and strongly encouraged Isabel to write and study on every topic she found interesting. In her pursuit of understanding personality, she worked to create a means of classification, which led her to the works of Carl Jung (and the idea of archetypes), whom she corresponded with greatly.

In time, Isabel grew up and married Clarence Myers who at the time was a law student. The personality differences between the two were so pronounced that Isabel sought to research the differences in personality, following in her mother's footsteps (which we'll get back to in a second).

Isabel and Clarence had their children which Isabel homeschooled until they went off to college. At the beginning of the Second World War, Isabel wanted to help with the war effort. She found that although there were many people volunteering for the war, a lot of them hated the jobs they had. Isabel thought that the differences between people should be understood and used to help them find happiness and fulfillment in whatever endeavor they did. She continued the development of a system to classify personalities. Isabel worked with her mother to further develop the theory and created a questionnaire that would help determine the best career fit for a person within the military. She was fascinated with the statistics and was seeing a lot of success with the results. It eventually became a standard form of the military, many colleges, universities and in business.

The system, which became known as the Myers-Briggs Type Indicator (MBTI), has become the basis of many new methods of profiling personality. Currently there are many different models and forks off the MBTI, but the underlying classification process is the same. There will be a questionnaire, which will put a person on a scale of between two opposite traits and within several categories (sometimes four categories, sometimes it's five). Then the combination of different traits will lead to a detailed analysis of a person with information on; how they focus their attention, acquire information, make decisions and orient themselves with the rest of the world.

It's important to use this tool for introspection and insight to oneself, as it helps achieve goals and meet one's potential. However, the aspect of the categorization that I want to highlight is how certain personality types address risk in comparison to others.

Let's look at the categories and how they relate to risk:

Personality characteristics and how they relate to risk

The combination of these characteristics creates 16 different personality archetypes. Each one will approach situations and risk in a different manner. Figure 4-1 shows the essence of each of the archetypes.

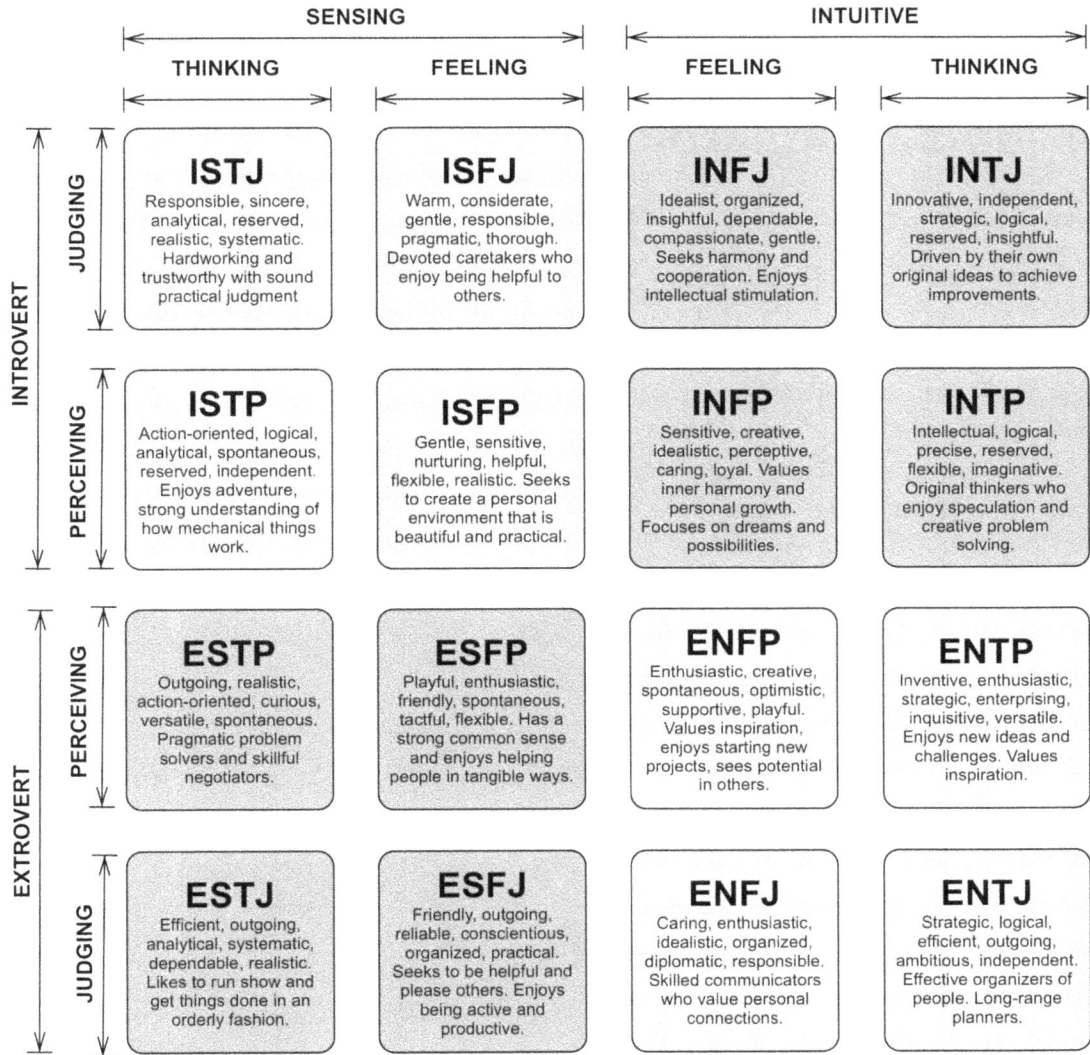

Figure 4-1. 16 personality archetypes

4.2 Understanding risk in personality types and circumstances within organizational roles

There have been studies that have shown that the amount of risk that a CEO is willing to undertake can be attributed to several things. Personality type, as we have seen earlier in this chapter, plays a big part in risk tolerance. Another personal attribute that may be taken into consideration (but is not usually discussed publicly) is that of marital status. Studies have shown that unmarried CEOs tend to invest more of an organizations percentage of assets into growth activities and capital expenditures, thereby accepting more risk. Normally this is not something that can officially be taken into consideration by a Board of Directors because it is discriminatory. However, this is not to say that the practice is not done.

(4) Water, urine, wine, or any other simple liquid you can get in reasonably large quantities will dilute gasoline fuel to a point where no combustion will occur in the cylinder and the engine will not move.

Figure 4-2. A tactic from the CIA's "Simple Sabotage Field Guide"

Senior leadership within an organization can get a lot of attention, but often, it is the employees and interpersonal interactions that set the tone for the company. The dynamics of an employee's personal life situation can also have far reaching effects on an organization. What is happening in one's own life can change the amount of risk tolerance someone is willing to accept, or the clarity in their decision-making processes.

Personal issues can affect all employees, the roles they occupy and the people they interact with. Though, the level to which it affects the organization will differ vastly between a member of the janitorial staff and a senior executive.

Large and successful organizations are well aware of the importance of ensuring that staff are provided with all the tools and resources required to meet their personal needs and achieve their career potential.

The scale of the potential for harm that can occur from the actions of staff vastly differs based on a person's role, but that does not mean that less damage will occur from a lower position. There is a declassified CIA document called the *Simple Sabotage Field Manual*. It provides insight into the many ways in which someone could cause maximum damage to an organization from within.

Some of the actions suggested required minimal access or station within a company such as: urinating in the gas tank of vehicles, or gathering large amounts of oily and combustible material and leaving them in the basement then "accidentally" lighting them on fire.

Other actions require an office setting with other people to perform the following:

1. Make "speeches". Talk as frequently as possible and at great length. Illustrate your "points" by long anecdotes and accounts of personal experiences.
2. When possible refer all matters to committees, for "further study and consideration". Attempt to make the committees as large as possible – never less than five.
3. To lower morale, and therefore productivity, be pleasant to inefficient workers. Give them undeserved promotions, discriminate against efficient workers and complain unjustly about their work.

There is great irony in this, because employees within organizations everywhere are unwittingly conducting many of these documented, covert and subversive tactics. An employee may functionally become a saboteur within an organization simply by being incompetent, inefficient or intentionally overbearing.

This brings us back to the discussion of personal life situations. If somebody has personal issues that they are dealing with, it may change their focus, attentiveness, drive and logic in decision making.

These periods are normally transient in nature and not states that define who people are in every situation. Elongated periods that affect people in a negative way will also affect everything they do and the people around them negatively. Support networks, through family, friends and peers can help individuals navigate their personal situations, but ultimately regarding organizational output, they will be evaluated on what they can do in their current state.

Personality however, is something that is not transient (unless a drastic change within someone's life stimulates a transformation). So, by knowing what personality type someone is and that of the people around them, the decisions they make can be calculated with greater accuracy than simply comparing historical trends or guessing. The relevance of this level of organizational introspection can be seen in the following example:

Kodak was the epitome of a company that strived for continual innovation in its core business. This brought them great success from the late 1800's onward, as they essentially had a monopoly on the film and camera industry for much of the twentieth century. As the world changes over time and companies must adapt, innovate and prioritize to stay relevant, there are pivotal points in the evolution of a company that can spell success or demise, depending on their choices and focus.

In 1986, Kodak created the first megapixel digital sensor and subsequently, the first digital SLR camera. It had the market share, the technology and the means to remain at the forefront of photography into the new millennium. What it did instead was to limit the use of the digital innovations and focus most of its resources on optimizing the chemical film processing technology. Kodak had prioritized continual improvement in a dying industry, and thus, died with it. Kodak's decline was not because it could not develop cutting edge technology, but because their leadership would not risk the change in focus of their core business in keeping with consumer demand and with the changes in technology. For them, it was simply too risky.

Google and Facebook have been so successful because of their ability to adapt and fundamentally change their business models to evolve and they not only continually enter new markets, but create them to stay at the forefront of change as well. Within the culture of these companies is a strong entrepreneurial spirit, where innovation is fostered and then made profitable. This is only possible by having teams that have that a particular mindset, certain skills and work well together.

Some people may be the smartest in their field, but that doesn't make them suited for an entrepreneurial environment. There are countless stories of entrepreneurs dropping out of school to pursue something that they were passionate about and achieving great success. Names like Steve Jobs, Mark Zuckerberg and Bill Gates are just a few of these. However, a successful company is never made up of a single personality type. You cannot have a room full of high functioning visionaries all agreeing on a single vision, nor can you have a room full of tenured professors come up with the next great startup. Success requires a balanced orchestra of creativity and stability to achieve sustainable greatness.

Some organizations like the pharmaceutical company Merck, have the concept of "kill fees". This is a reward given in stock options, to scientists that abandon failing projects. It saves money for Merck and it allows for the re-organization of staff to work on successful projects, thus generating more revenue. Incentives like kill fees help bridge the gap between evolutionary and revolutionary genius.

There are many ways in which risks are taken by individuals: a personal financial risk, a risk of job stability or credibility, financial risk to the organization, a risk to a client, or the public, a risk of litigation and a risk of safety. Some risks can be calculated and beneficial, while others may be negligent or irresponsible.

People that are given the opportunity to pursue something that they are passionate about will invest more time, effort and drive into their work. Intrinsic motivation is a very powerful motivator, which can be directed to the benefit of a company by giving employees the freedom to spend time on their own projects. Companies like 3M and Google have a significant percentage of time allocated to projects that staff have chosen themselves. Not all projects will add revenue to the company's bottom line, but many might. It is a proven gamble with high payoffs and low risks. Other people prefer extrinsic motivation by way of larger salaries, bonuses, or recognition. Properly matching people to projects and providing motivation that best fits their needs will reduce the project risk and allow for greater autonomy and less management overhead.

Leaders within all levels of an organization are often driven by fear of failure or the desire for success. Those that have a proven track record of success may wonder how long they can sustain it. The longer that a leader is outstanding in their performance, the greater the chance they will feel the pressure to stay on top. This may force them to become more conservative in their actions to reduce the risk of failure. Ironically, this counters the process that they took to get to that position in the first place and it becomes a self-fulfilling prophecy.

Leaders inherently need to take risks, and the good ones will weigh out the possible outcomes of their decisions, as well as know the difference between being calculating and being reckless.

4.3 IT Budget as a variable expenditure

When designing a solution, the budget is one of the three pillars of success for the initiative. The other two are quality and provisioning time or time to market. Or to put it in simpler terms; fast, cheap and good. This is usually represented in a Venn diagram with the tag line "pick two".

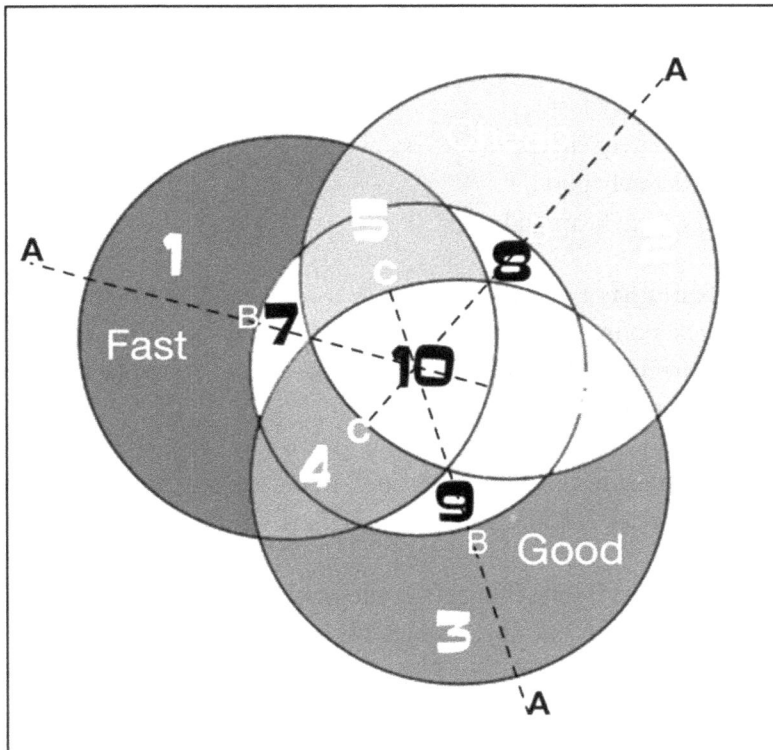

Figure 4-3. Good, fast, cheap, Venn diagram

While witty and with some truth, it is not the whole picture.

If we take that same Venn diagram to the next level, we can drill down a bit more.

In this diagram, we have three large circles which depict the three business directives: fast, cheap and good.

Then there is a fourth, smaller circle in the center that intersects all three large circles. This smaller circle is the zone of viability.

The intersections of all these circles divide the diagram into other zones. You will see the letters A, B, C across the diameter of the large circles. This represents the degree to which something is fast, cheap or good. A is the minimal amount; B is moderate and C is high.

1. This area is not very fast
2. This area is not very cheap
3. This area is not very good.
4. This area is very fast and very good.
5. This area is very cheap and very fast.
6. This area is very cheap and very good.
7. This area is moderately fast.
8. This area is moderately cheap.
9. This area is moderately good.
10. This area is very fast cheap and good.

Now let's say that each area has a rating of 1 to 10; A is 1, B is 5, C is 10. Then instead of the "pick two" phrase, let say "pick a number from 3 to 30". This number represents the total of the weights placed in each of the categories. Let's call it the resource value. Now place the values in each of the regions to total that number.

For example, if we choose the number 15, then put 5 in cheap, 5 in fast, and 5 in good. Based on these values, I end up in zones 7, 8 and 9. All of which are within the zone of viability.

If we take the same number (15), but change the allocation, let's see what happens. This time let's place 7 in cheap, 3 in good, and 5 in fast. So now we end up in zones 5 and 3. Two of the values (7 and 5) end us up in the zone of viability.

If we have less than 15, no matter what combination we use, we will only have one value in the zone of viability. For an initiative to be successful, it needs to have at least two values in the zone of viability. If an initiative has a resource value of 18 or higher, then it can reside in zone 10, also known as the unicorn zone.

So far, this is all academic with pretty colors and a Reuleaux triangle, so how does it become relevant? If you determine the minimum and maximum possible capabilities organizationally, then you can map them to your initiative capabilities. For example:

The fictitious company, Ronnie's Hamburgers has determined that with the best equipment, fastest people and finest fresh gourmet ingredients, it can make a luxury burger from order to exit in 30 seconds. The total cost for Ronnie's to make the burger is $10 per burger.

After operating expenses, Ronnie's has determined that to be minimally profitable it needs to make about $30,000 gross per day. So, that would be a minimum of 2000 customers buying a single burger for $25 each. The profit would be $15 per burger, or $30,000 per day.

This scenario is a 21/30 scenario, fast = 10, good =10, cheap = 1.

Now let's say that the shareholders at Ronnie's want a better return than 0.00001% on their investment. However, quality and speed of production cannot be compromised. This means that the resource value needs to be increased from 21 to something higher. Since the optimization on good and fast, cannot be increased, then the operating costs need to be reduced for the resource value to increase.

What would be required to go from a 1 to a 10 on the cheap scale? Let's say that everyone works for free, there are no operational costs and no costs for the ingredients. Now Ronnie's has 30/30 for resource value and is riding a unicorn (zone) into the sunset as the shareholders swim in their money.

Now let's come full circle and see how this applies to IT infrastructure. The IT budget of an organization is often viewed as an expense sinkhole, just like the ingredients for gourmet burgers at Ronnie's. For the organization to get closer to the unicorn zone, it often cuts budget from areas that feels it can, such as IT.

The risk, besides the obvious ramifications, is that the viability of ongoing projects may change based on the budget provided to IT. If the budget is cut, then other areas need to be bolstered to maintain the minimum viability zone of 15/30 for ongoing initiatives. The consequence of not knowing the exact effects of budget cuts could be disastrous for initiatives down the road. Again, the fail-fast methodology is used as an immediate fix. Waiting until later often yields heavy handed fixes, such as layoffs and closing of departments. That then has cascading effects like employee exodus and mass stock sell-offs.

When planning an IT strategy, the design of solutions should be logical before physical, so that if during the initiative lifecycle the budget changes, then alternate physical plans can be put into place to compensate. It is also a good idea to have multiple logical plans (primary and backup), so you know what can be adjusted based on physical resources, risk and budget.

4.4 Defining a dynamic and adaptable IT culture

The culture in a company is what binds the people in it together. It provides a story of the identity of the organization, the reason it is there and the history of how that happened. It has rituals, values and an internal language that define it from other organizations.

OUR MESSAGE IS SIMPLE:

Where our music is welcome,
we will play it loud.

Where our music is challenged,
we will play it louder.

Figure 4-4. Culture clashes can occur if not managed and directed

It has been said, "Organizational culture is civilization in the workplace" –Alan Adler. It is a microcosm of society with hand-picked citizens; a type of social experiment. The values, or tenets of an organization are re-enforced by behavioral norms, whether they are defined and spoken. Actions speak louder than welcome packages and corporate policies. The longer that a corporate culture has been set (by either deliberation or circumstance), the harder it is for it to change.

As a rule, if you assimilate new people into an organization that has similar values to those that are being strived for, then the process becomes a familiarization exercise versus an attempted indoctrination. Because of this, acquisitions and mergers are generally very difficult to navigate. Different companies have different values and corporate cultures. Thus, if merging entities are too dissimilar, then there will be a huge attrition rate as people flock to leave in droves due to the change itself, the fear or feelings of uncertainty and the perceived lack of long-term stability.

Some organizations have a positive culture that is accepting, dynamic and conducive to change. These are the ones that evolve with the times and reflect the goodwill that is put into it by employees by giving more allowances and freedom to achieve the corporate goals as they see fit. Other organizations have a more structured approach that clearly defines and measures what is expected, and accepted, and what is not. Neither approach is "better" per-se; they just have different desired outcomes and methods. However, you cannot take someone who is successful in one environment and automatically expect him or her to be successful in the other.

Joe Strummer, the lyricist and lead vocalist of the punk band "The Clash" once said "Where our music is welcome, we will play it loud. Where our music is challenged, we will play it louder." This is the same for all strong organizational cultures. If a culture is well established and rigid, it will be difficult to change. Or if it is relatively new, but embodied by fervent and passionate people, then it too will be difficult to change.

Culture differences are one of the prime antagonists to effectively integrating different companies (or even regions or departments within a company). The problem that companies face, especially those in diverse communities, or spanning large geographic regions, is how to maintain a unified culture, while respecting the multiple regional cultures. Differences in the way subsections of an organization operate can make cooperation and decision-making a very tedious task. There must be a balance, with a bit of give and take, to achieve a "unifying culture".

Identifying with a corporate culture and with its idiosyncrasies is often difficult to achieve by the same group of people. This is because many tenets, practices and behavioral norms are so implicit and ingrained, that it is difficult to see them in oneself. A parallel of this is the recognition of accents and dialects in language. Unless you take an active interest in seeing the differences and similarities, you may not be able to point out the specifics of your own.

Here is a conversation between two guys, one from Newfoundland and one from South Wales, both with different dialects of English:

NFLD: *How's you gettin' on, cocky?*

WALES: *Stop chopsing and buzzing! I'm not being funny. I'm under the doctor.*

NFLD: *Oh me nerves! G'wan b'y! Your some crooked eh? Last night you was tits up in the rhubarb.*

Translated into generalized North American English:

NFLD: How are you today?

WALES: Stop bugging me! I'm being serious. I'm not feeling well.

NFLD: Oh, come on! You're unjustifiably grumpy today. If I remember correctly, you were drinking quite a lot last night.

In both translations, you can see that there is a localized version of English based on the region. In Newfoundland English, the word order is inverted and there are several abbreviations and metaphors; whereas in Welsh English there are common sayings and similes.

The languages are both English, but they are different enough that not everything translates properly without an understanding of the culture. The same is true of cultures within work environments. Some things just don't translate and you need an understanding of the people, the inter-personal dynamics and thought processes to understand and appreciate the intricacies.

Here is a conversation between two guys, one from the technical pre-sales department and one from the technical delivery / operations department.

PSALES: Hey, you got a second?

DELOPS: Sure, I only have a moment though.

PSALES: We sold client X, Solution Y and just need a bit of your expertise to help bring it across the finish line.

DELOPS: I thought we weren't going to sell solution Y until we did some validation in the lab first?

Translated into generalized North American English:

PSALES: Hey, I'm going to involve you in a new project that we sold and now must deliver in a couple of weeks, with no scope, requirements or direction from the client yet.

DELOPS: I know you are up to your tricks and your tone dictates it's a doozy. This will be good for a laugh, but more of a crying on the inside kind of laugh, because I know I'll be dragged into a mess with little regard for the other projects that I have on the go, my general lack of sleep or my sanity.

PSALES: The client has the hardware we sold them, but has no idea what it is or what it should do. We're hoping that you can provide some sort of technical justification for it and make them feel good about it, then deploy it and train them.

DELOPS: I can make the impossible happen again, but I have a list of demands starting with coffee, a new lab environment, some pizza and some respect while I fix the mess you've made.

In the translations, you can see the subtext in the exchanges and a thinly veiled tone of derision. These two people do not work well together, but there's a good reason for that, which is based on non-stated assumptions.

Technical pre-sales did not mention that this is the 3rd project of this type with the client and that the client preferred it this way so it is a more collaborative process while working on a tight schedule.

Delivery/Operations did not express that they want a scoped and validated project because they care about the quality of the work and because they feel it reflects on them personally.

So, the delivery engineer feels like they are required to come to the rescue of the organization and the sales guy is wondering why the engineer is getting all riled up over a simple project that's been done several times before. Without the backgrounds from each person, the communication is much more silent and the unspoken interaction is lost, leaving only resentment and misinformation.

Mergers within companies are normally underpinned by financial calculations and market motivations. Their success however, is strongly influenced by how well the cultures can adapt and integrate with one another. If the decision-making and leadership styles are vastly different, then it can delay important decisions and create turmoil and turnover with employees. Cultural assumptions from the former separate companies will not always translate similarly under the umbrella of the new company.

An example is a difference of having rock-star employees that act in a solo manner and achieve high revenue and quick (but volatile) results, versus having a team that works well together over a long term for more overall revenue and reliable forecasting.

Each company will have its own individual scenarios that must be addressed, but there are some common risks that can be addressed by doing the following:

1. Define and socialize a shared decision-making process that works for all parties.
2. Create an internal identity, or brand within the organization that resonates with employees.
3. Understand the compensation programs from each former company and present a new common program that is beneficial to all staff. This is especially important when a hostile takeover occurs.
4. Ensure that the most valuable and revenue generating parts of the former companies continue to work efficiently throughout the merger process.
5. Identify the strengths of both existing cultures and incorporate the best parts of each.

4.5 Elastic infrastructure and cloud services

Writing about cloud services in just a sub-section of a chapter does not do it justice. One could write volumes about the evolution of computing, the origin of modern day service offerings, or the future of computing. For the purposes of this book, let's just sum it up by saying it's awesome. Or, OSSM.

On demand
Scalable
Self-service
Measurable

On demand means that you can get the resources you need for your workloads without the initial capital investment. Scalable means that it can grow or shrink in a consistent manner without any dramatic change to processes or infrastructure. Self-service means that the team managing the infrastructure can do so without the involvement of vendor distribution, build and staging teams, facilities teams, network, provisioning, etc. This reduces operational overhead and provisioning time.

There is an incredible wealth of interesting things that can be with cloud services such as:

1. Spanning multiple locations with layer 2 networking by leveraging software-defined networking.
2. Building clustered storage systems that can scale performance tiers based on active workload.
3. Offloading work from traditional physical devices, like load balancing, security services or VPN connectivity.
4. Having drag and drop deployment of application services, operating system templates, enterprise desktops, relational / non-relational databases.

The list of things that you can do in the cloud can go on and on. However, the current conversations at the coffee machine around cloud are no long about what you can or cannot do; that's been covered. Rather, the conversation has become: what are the risks and caveats of implementing cloud technologies? There is always a cost / benefit calculation that's involved when moving to cloud services.

Considerations you need to address are: data locality, ingress and egress traffic costs, management access security, hybrid environment considerations, SLA and performance metric tracking. Many enterprises these days are "all-in" when it comes to cloud services, therefore the conversations are less about migrations and proof of concepts, and more about operational concerns.

For the organizations that have been slow or cautious on the uptake of cloud services, they are now finding that the sheer number of options can be daunting. Specific skill-sets need to be obtained, either through professional services and managed services or internal training.

The staff that manages traditional silos of technology cannot confidently take ownership of a cloud initiative without a dramatic shift in their mindset and skillsets. This is not to say that it is impossible, but that the potential for success and optimization is lessened and ends up costing the organization more money due to inefficiencies.

4.6 Technology re-use

Hardware manufacturers are often not given enough credit for the products they build. One reason for that is planned obsolescence and imposed hardware lifecycles. It is often the recommended practice to put lifecycles on hardware, so that every 3-5 years there is hardware refresh regardless of capabilities or performance factors. This is both a good thing and a bad thing.

First off, it ensures that all the latest capabilities and enhancements in the newer hardware have the possibility to be taken to its fullest advantage. It also renews the warranty on the production equipment, reducing the risk of component failure related downtime and out of warranty expenses.

The downside is that 3-5 years is often a fraction of the lifespan of higher end enterprise equipment. What happens to this perfectly good working hardware? It gets refurbished and sold a second time. Refurbishing companies will buy all the end of lifecycle hardware from large companies for pennies on the dollar, then sell it right away to a third party for 5-10 times what they paid for it. The large company gets some money back and the inventory off their books. The refurbishing company makes a ton of cash as the middleman. And the third-party company gets a great deal on perfectly good (and burned in) working enterprise equipment.

The only thing to note is that the hardware may be out of warranty from the manufacturer, which is a risk. However, because of the reduced price of the equipment, redundant components can be purchased for standby replacement rather inexpensively.

A good practice is to try to get hardware that still has current firmware available for it. This will ensure that security and bug patching is not neglected. If a product is outside its vendor support lifecycle, you may not be able to get any more firmware updates. In that case, you would need to evaluate whether there are any documented vulnerabilities or risks associated with the hardware that are not acceptable risks.

Manufacturers will list the hardware support lifecycle on their websites. The phases would be the latest version, in support and out of support. Some manufactures make their own hardware, while others simply make the software and use a common OEM (white labeled) platform to run it on. The secret sauce is in the software. Sometimes that secret sauce isn't very good or the company that makes it doesn't market it very well. The underlying hardware may still be good have some life in it and if you can get it for a very low price, it may be worth the risk.

The Symantec 5420 Security Gateway is an excellent example of this. It was essentially an x86 server with a Celeron CPU and six 1Gbps NICs in a 1U form factor. It did not have a wide market acceptance and was eventually discontinued.

However, by installing PFSense (which is a derivative of FreeBSD customized specifically to act as a firewall), you have an open-source enterprise-class firewall that is incredibly secure, robust and has an active 3rd party plugin system. Since it is an x86 platform, just about any version of Linux, Unix, or even Windows could run on it (assuming driver were available).

This was not true for many network devices that are built around custom ASIC processors. However, about a decade ago, Linksys started using open source software as the base for their firmware. Because of this, a lawsuit ensued and they had to release the code they used on their equipment. From this code came a whole slew of new open source router firmware with many advanced features. These are the WRT derivatives, most notably of which are DD-WRT and OpenWRT. Consequently, these firmware derivatives can be installed on more than 1000 devices that have a similar ASIC hardware base.

One strategy for building robust architectures with cheap and aging hardware is to build a high level of redundancy into the environment using commodity equipment that can be easily replaced. In situations like these, the key to minimizing the work required for repair and overall downtime is configuration management, fault detection and automation.

So far, we have been talking mostly about networking equipment. When it comes to data and applications, it's a bit of a different approach. The capabilities of processors have increased exponentially and thus the way that computing has progressed has been towards virtualization and micro segmentation of services. This has given us huge consolidation ratios and reduction of resources required (like space and power) to perform the same job function. The problem we encounter is that by using older hardware to do the same function, we simply can't go back to doing everything we would expect on new hardware. If we did, then the processing inefficiencies would force us to use much more equipment and negate any savings that would have been garnered from using old hardware in the first place.

We must look at it from a computing resource perspective: how much aggregate CPU, RAM, storage, etc., do we have available with our re-used equipment? How much power, space, and redundancy do we require to ensure that the environment will be resilient and reliable enough to perform the desired functions? What are the largest risks with the equipment that is being used, and how can we mitigate them?

Now that we've determined the total resources we have, let's look at TCO to see what the numbers looks like for capital and operational expenses. Then let's look at what services we would like to offer within the environment and map them out logically. With the capacity available, can we offer the same services with the old hardware that we can with new? Can we modify the underlying architecture so that we can provide the exact same services with the old hardware? Where possible, can we leverage core services vs. bloated services to provide the same desired solution? What are the opportunities for optimization? What about from the desktop side of things? Can we derive benefit from technology re-use there as well?

The desktop side is probably one of the easiest and most cost-effective ways to start a pilot project for re-use. First, let's classify the various types of desktop users; task workers, knowledge workers and power users. These types of users determine the processing power and RAM required to provide the desired desktop solution.

Desktops have traditionally been part of client server models, with storage and collaboration provided by servers and application processing provided by the local desktops. This model has changed over the years as many applications are being provided by web services and desktops simply provide a local GUI and file access. A perfect example of this is the popularity of the Google Chromebook. This laptop has lower than normal laptop specifications, is very inexpensive and provides access to web services only. The performance is good; the experience is great and there is virtually no task that cannot be done on it for most scenarios.

By taking a chapter from the Chromebook model, you can start providing enterprise applications and services through web services, session based computing and VDI, instead of running them locally. This will reduce the overall cost, configuration and management required for desktop installations. Over time, desktops can be replaced by solid-state thin clients and zero clients that have longer effective lifespans than traditional desktops. If capital purchases of new thin clients are not possible, then old desktops can be converted into thin clients by creating a simplified desktop interface with access to the required applications for business. Four common methods for doing this are:

1. PXE boot LiveUSB

By using a PXE boot environment with a lightweight OS distribution on a LiveUSB like Xubuntu or a micro distribution like Puppy Linux, you can provide a stateless desktop environment with very little overhead.

2. Session-Based Computing (Windows)

By using either a PXE boot environment to deploy the thin client OS, or using an embedded thin client on a bootable USB, or internally on a hard drive, you can provide a full Windows-based desktop environment by having users access a session on a Windows RDS host or pool of hosts.

3. Session Based Computing (Linux)

By using either a PXE boot environment to deploy the thin client OS, or using an embedded thin client on a bootable USB, or internally on a hard drive, you can provide a Full Linux based desktop environment by having a session on an LTSP host or pool of hosts.

4. Virtual Desktop Infrastructure

By using the same method as session based computing, you can change the client from a RDS client to a VMware Horizon View client. This will provide you with remote access to a virtual machine running either a Windows or Linux based desktop. The difference over a session-based solution is the granularity of performance you can provide per desktop and the ease of management.

4.7 DevOps culture

In 2008 Patrick Debois and Andrew Shafer started a discourse in the IT community around "Agile Infrastructure", from which the term DevOps was coined. In 2009, a series of conferences called "DevOps Days" started emerging in many places worldwide and thus the revolution started.

DevOps can be thought of in the simplest manner as having your infrastructure treated like code, hence the merging of Dev from Developer and Ops from Operations. By combining tools such as configuration management systems, automated deployment and provisioning systems, monitoring and ticketing, it is possible to create a highly controlled and scalable environment.

When looked at separately, the concepts are not new, but together they change the model of traditional IT significantly. Some of the benefits that organizations see from using DevOps are:

- Version controlling of infrastructure and subsystem builds.
- Lower failure rate
- Faster mean time to recovery
- Shortened time to market
- Automating infrastructure changes

Organizations that have embraced DevOps have also embraced the highly collaborative culture that comes along with it. That is the biggest difference from traditional IT, collaboration and breaking down of silos. The very nature of the culture requires both Developers and Administrators to take an active role in understanding and contributing to the betterment of the infrastructure on a frequent basis. In traditional IT, changes to an environment are done over the course of yearly or multi-year capital projects that rely on a budget. If the budget is not sufficient, then the infrastructure may stay in a holding pattern until the hardware or software is considered end of life and a forced expenditure allowed for an upgrade project.

The use of Agile processes allows for multiple concurrent projects to occur with rapid iteration and incremental gains. This is in stark contrast to traditional IT, which requires a waterfall approach such that one task or project builds upon the previous one. A waterfall approach has a lot more dependencies and takes longer but has its merits. The biggest complaint with the waterfall approach is what is referred to as analysis paralysis. This is when you spend so much time thinking about and analyzing a design that nothing gets done.

On the other hand, Agile processes limit the opportunities for in-depth planning for an environment. Rather, it addresses issues by observing them and iterating to the next version with a fix or retooling of the problem area. The scope of any issue that may occur is reduced by the limited scope of the provisioning process while entering the dev/test phase.

There are benefits of using both agile and waterfall processes when building an environment. Dependencies like time to market, recoverability, availability of dev resources and internal IT willingness for change, will determine how much of each model can be adopted and where they are used.

There are several ways that DevOps can be adopted, but the most accepted seems to be incremental adoption using the "Three Ways Principle". Gene Kim made this popular in his landmark book named "The Phoenix Project". They are listed below:

The First Way: Systems Thinking

The focus is on the performance of the overall IT environment as it relates to the output of the business as a whole. This ensures that silos are not optimized in such a way as to negatively affect other parts of the organization. Systems Thinking puts the following concepts into practice:

1. Never pass a known defect to downstream work centers.
2. Never allow local optimization to create global degradation.
3. Always seek to increase flow.
4. Always seek to achieve profound understanding of the system.

The Second Way: Amplify Feedback Loops

The focus is on getting feedback from all groups involved in the business processes. This is done by creating greater communication within the organization, or getting feedback from and responding to customers. Amplified feedback Loops puts following concepts into practice:

1. Enhance communication internally.
2. Get feedback from customers and respond to them. Create a continual dialog.
3. Provide the feedback and knowledge gathered to the people that need it and make it accessible within the organization.

The Third Way: Culture of Continual Experimentation and Learning

The main two points of this way are experimentation and practice within the work culture. It is important to take risks and try new things, learn iteratively and develop a mastery of skillsets. Creating a Culture of Continual Experimentation and Learning puts the following concepts into practice:

1. Set aside time to improve daily work and mastery of skills.
2. Create rituals that reward teams for taking risks and experimenting.
3. Introduce faults into the system to increase resilience.

4.8 Open Source Software for Desktops, Servers and Line of Business Applications

From the 1950's to the 1970's, it was standard that all operating software, compilers and source code was provided with a hardware purchase. This was mainly because, to run the software on different hardware, many things in the code had to be tweaked and modified to make it work.

In the 1970's and 1980's, it started to become more common for software vendors to license their software and to restrict access to the code as complexity increased. However, in academic institutions, any software installed in the computer labs had to have the source code provided in order to ensure that no malicious code or backdoors were embedded.

In the mid 1980's Richard Stallman created the GNU project, to create a free and open source operating system. This started what was to become the beginning of the FOSS (free open source software) movement. In 1991, Linus Torvalds released the first version of the Linux kernel, which started a revolution.

The 1990's were a time of the dot-com boom; every day millions of new people were getting online for the first time. Websites were popping up faster than you can imagine and it was the wild west of the Internet. The open source Apache Webserver became the de facto standard, surpassing all other webservers by orders of magnitude in market share and ensuring that the FOSS movement secured its place in the future of computing.

Nowadays, many large companies run open source software for their primary line of business software, their backend servers and their desktops. Massive companies like Google, Facebook, Amazon and many more, have been using primarily open source software on an unprecedented scale for years.

Desktops however, have long been the domain of Microsoft Windows because the applications that would run on them were not available on Linux or other OS platforms. This has changed more and more over the years as the model of computing architecture has also changed. Many desktop and line of business applications now are available as a web service, hosted either internally or as a SaaS offering, such that the only "application" that is of any importance is that of a web browser, most of which are now multi-platform.

The Windows-only web browser Internet Explorer "IE" (now Edge) was the dominant browser with nearly 70% of the market share in 2009. This meant that websites and services would be specifically designed for IE and not function properly or at all in other browsers. Since then, the decline of the market share of IE was then inversely proportional to the rise of the market share of the Chrome web browser from Google. In mid-2012, they were both tied for 30% each of the global market share. The trend has continued and as of 2016, Chrome had more than 60% and was still rising.

This conversion to Chrome meant that the dominant browser that was once a single platform application has now become multi-platform. The Chrome browser is available on Windows, MacOSX, Linux, Android, Apple iOS and many other platforms. That fact, combined with the move to web based applications for business has dramatically changed the requirements of a normal desktop environment.

Google has taken full advantage of this change by releasing the Chromebook laptop specification, which allows hardware vendors to avoid operating system licensing fees per unit when selling their laptops. Chromebooks run ChromeOS, which is a slimmed down graphical Linux distribution. Most applications run directly from the chrome web browser with very little configuration. In 2016, the ChromeOS added support for the full library of over 2 million Android applications, making it a cross between a laptop and a tablet further enabling choice and flexibility.

If we look away from the home and educational market and focus back on the enterprise market, and the desktop platforms that are used therein, who can we look to as converts to an all Linux desktop environment? These six high profile organizations have all converted to a Linux desktop environment:

1. Google

No surprise there, but what is interesting is that internally they are not using ChromeOS or Android, but rather a fork of Ubuntu which is called Goobuntu. Tens of thousands of Googlers across all departments and levels are using it as their primary OS.

2. CERN

The famous nuclear research laboratory has more than 3000 desktops running a combination of Scientific Linux, RedHat Enterprise Linux, Ubuntu and CentOS.

3. NASA

Remarkably, Windows XP used to be the desktop used on the International Space Station. They changed over to Debian in 2012. Anecdotal accounts also indicate a mix of RHEL for desktops that need enterprise support. CentOS and Ubuntu are also used on many other machines.

4. The US Department of Defense

The DoD developed a small distribution called LPS (Lightweight Portable Security) that boots of a USB drive in a stateless high security desktop environment. It can be plugged into any computer and used in non-secure environments leaving no traces.

5. The City of Munich, Germany

The city migrated all their desktops to a distribution fork of Ubuntu (called LiMux) from Windows, over the course of a decade (from 2003 to 2013). They now have over 17,000 desktops running LiMux, which saved them tens of millions of dollars in licensing.

6. The French Gendarmerie

The military police force in France migrated 90,000 desktops from Windows to Linux over the course of a decade. They are using a fork of Ubuntu called Gendbuntu.

For applications that cannot be "webified" for cross-platform compatibility and must run on the Windows OS, there are still four options:

a) Get the software vendor to port it to Linux.

This is an uphill battle and will not be possible 95% of the time, if it does not make financial sense for the vendor.

b) Run the windows application in an emulated layer.

The WINE (WINdows Emulator) project allows for this and it supports thousands of applications. For enterprise support and greater compatibility, a company called Codeweavers makes a commercial version of WINE called CrossOver. Walt-Disney Feature Animation moved to an all-Linux desktop environment and used Crossover for Photoshop.

c) Run the Windows application in a slim stateless Windows VM, and present it in a seamless window.

VMware makes a product called Workstation for Linux that allows a Windows VM to be run in the background and specific applications installed on it to be presented as native applications on the desktop.

d) Publish applications to an RDS server or VDI solution.

Microsoft has session-based desktop server software called Remote Desktop Services. With it, you can publish a specific application to a group of remote users on multiple operating systems, including Linux. It can have a seamless window presentation, so to the user it seems like a native application. VMware Horizon View provides an excellent management platform for this type of solution.

4.9 Continuous Integration / Continuous Development

A subset of the DevOps method is the feedback loop of continual improvement. This is done by; experimenting, learning, testing and releasing updates to the infrastructure. This process is called CI/CD, or continuous integration / continuous development.

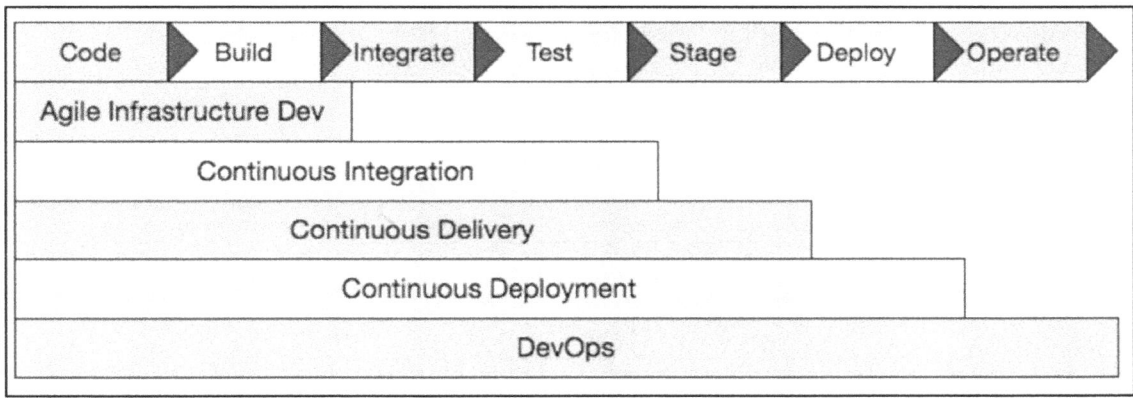

Figure 4-6. All the phases within the DevOps cycle

When the infrastructure is treated as code, this method ensures that it will continuously improve over time. Some software that can help do this is VMware vRealize Automation (vRA), Chef, Puppet, Ansible and Saltstack. Provisioning automation and management of server lifecycles is vital in the CI/CD processes. Configuration management systems ensure that the infrastructure code is identical when deployed in each environment, whether its dev, test, staging or production.

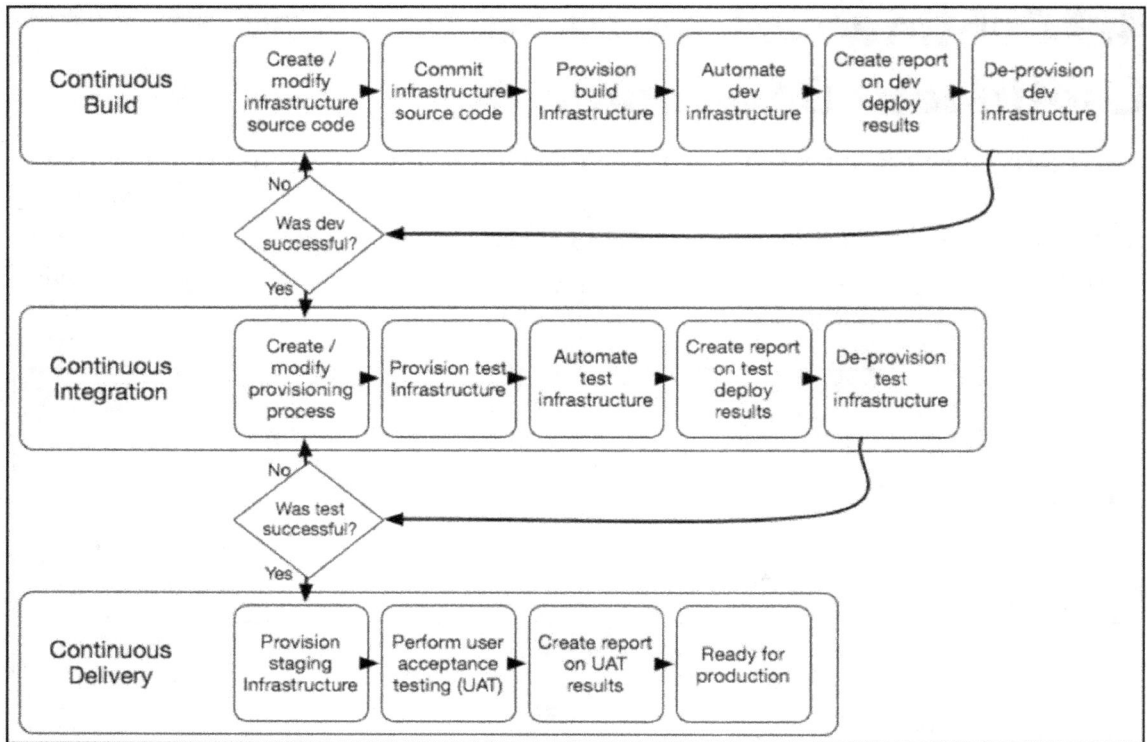

Figure 4-7. CI/CD workflow

4.10 The risk of tribal knowledge

The term "guru" is used historically when referring to a spiritual leader, a teacher, a counselor, or one whom relays the experience of a lifetime to help mold and train a student. However, in IT you often hear it from the less tech-savvy people about someone whom just setup their printer or removed a slew of viruses from their computer.

The etymology of the word is interesting, as it can be broken down into its parts. "Gu" means darkness and "Ru" means one who dispels it. So, a guru would be one that dispels the darkness of ignorance.

Figure 4-8. A Unix admin contemplates training a new hire

Gurus, shamans and leaders in many cultures passed certain knowledge through the generations in a strictly oral fashion only. Sometimes it would be passed down in in legends and stories, or other times in song and ceremonies. Ancestral knowledge was often very protected, sacred and only given to those that were "worthy" of it.

During the time of the Roman Empire, the Romans found it extremely difficult to control the Celts and subjugate them because of their lack of understanding of the Celtic culture, beliefs, hierarchy, religion, etc. This was because the Celts had a predominately oral culture and the record of their history was not in the written form and thus, to a certain extent, it offered protection to the Celts.

Fast forward a couple of millennia to modern times and let's look at the parallels in IT. Individuals and teams have technologies they manage, processes they adhere to and caveats of which they know. There are work-arounds, fixes, kluges, roadmaps, task lists and many other moving parts within an organization. Much of this knowledge is not socialized with those that are not deemed "worthy". It also tends to be the most valuable bits of intellectual property within the organization. This knowledge is what needs to be known to keep the lights on and the revenue coming in.

This knowledge that is known by a select few, is referred to as "tribal knowledge". It is acquired over many years from working in the trenches and learning what needs to be done to keep the infrastructure well-oiled and running smoothly. In this day and age when everything is digital, why would there still be this oral "pre-historic" (non-written) nature within a modern organization? Well, there are several reasons:

1. Job security.

If an employee feels that they are replaceable or under-appreciated, then they will hoard information related to their job role. This provides them with a greater feeling of security as it makes replacement of their position much more difficult for management.

2. Lack of organizational oversight and workflow.

If there is no requirement in place for the documentation of process, or even a review of the existing methods, workflows and strategy, then it will not happen. Employees will generally not create more work for themselves if it is not seen as a requirement, or mandated as a general practice.

3. Limited staff resources, lack of dedicated time for documentation

If staff are overworked, then the knowledge that is gained and processes that are used cannot be documented, because there is no time for it and they may have no motivation to do it. Organizational practices like documentation will reduce the per-person productivity, but it has the benefit of socializing tribal knowledge. This, in turn, allows more people to complete tasks in an identical manner and allow for scaling of the organization. It also allows for the ability to have KPIs (key performance indicators) to measure the performance of each individual employee against the defined processes and standards.

4. Employee churn and lack of a comprehensive onboarding and mentoring program.

By having procedures, workflow and roles / duties well defined, new employees have a much quicker ramp to effective utilization. If there is a knowledgebase available for common questions and problems with a mentor available to show them the ropes and to guide them, then new employees can feel more empowered and less like lost puppies or outcasts. A subsequent risk of having a high churn is that the people with the tribal knowledge are less likely to share it, because they don't trust that the new hire will be there for any length of time to use the knowledge or to pass it on successively.

The big risk with tribal knowledge is that when the people that possess the critical information leave, it will be lost and the business will not be able to replicate the work or processes that were done. This can affect the ability of the business to execute certain tasks, or to deliver products or services. It ultimately affects the bottom line of the company in a negative manner. If the organization has SLAs with customers, then it can be doubly negative as there may be financial or legal penalties associated with not being able to deliver services. Some things that can be done to protect against such risks are:

1. **Determine the maturity of operational processes.**

Does the organization have a standardized approach for common processes? What checks are in place to ensure that they are adhered to? Sometimes processes may be defined, but they are circumvented because they are complicated or counter-intuitive. Staff may even spend more time finding work-arounds, than following processes.

2. **Select a process or workflow and map it out.**

Choose a process and walk through it. Really understand all parts of it; who it involves, the dependencies, the risks, the costs, etc. Map it out on a whiteboard or mind-map so you can see the bottlenecks or areas for optimization. Once you have it defined, then get it reviewed by a focus committee that can provide relevant feedback for the specific process. Define a method to measure the performance of this process, be it time, adherence to a standard, quality, etc. Once finalized, document it, socialize it and then move on to the next process. Wash, rinse and repeat. Do this first for all areas that are "black box" or esoteric, then move on to the regular stuff like onboarding afterward.

3. Automate the process or workflow.

Work on continual improvement by analyzing KPIs. Review how staff perform each process and identify where the difficulties present themselves. Encourage feedback from staff and find methods to improve the end-to-end effectiveness of each process. Can it be removed, replaced or optimized? Is it redundant, or does it require a dependency that might be a risk? These are all factors to consider in the automation process.

Once an organization has embarked down the path of documenting and analyzing process, workflow and knowledge, then it can be socialized it with the rest of the company as well as with new hires. This creates a standard that is repeatable, measurable and can be continually improved.

4.11 IT agility measurement and risks

So, what does having IT agility mean? On the surface, it sounds like a buzzword, or rhetoric, like "cultivating opportunities for synergistic cross-pollination". It is not.

IT agility is not agile IT. Agile IT refers to the "Agile" methodology, as was discussed in the DevOps and CI/CD sections. They have very different definitions and although sometimes one may enable the other, they are not the same.

IT agility refers to an organization's ability to respond to opportunities and threats by aligning infrastructure with strategy in a dynamic manner.

The IT infrastructure that was designed in the first few generations of the client-server computing model helped enable business automation and the move to the paperless office trend. However, currently this same architecture impedes businesses from having agility and from adapting quickly to market changes, seizing opportunities or from responding to threats.

An example of this is monolithic single protocol storage systems. When most SANs were either direct attached SCSI or fiber channel, it would be quite a process to provide resources to individual projects or initiatives.

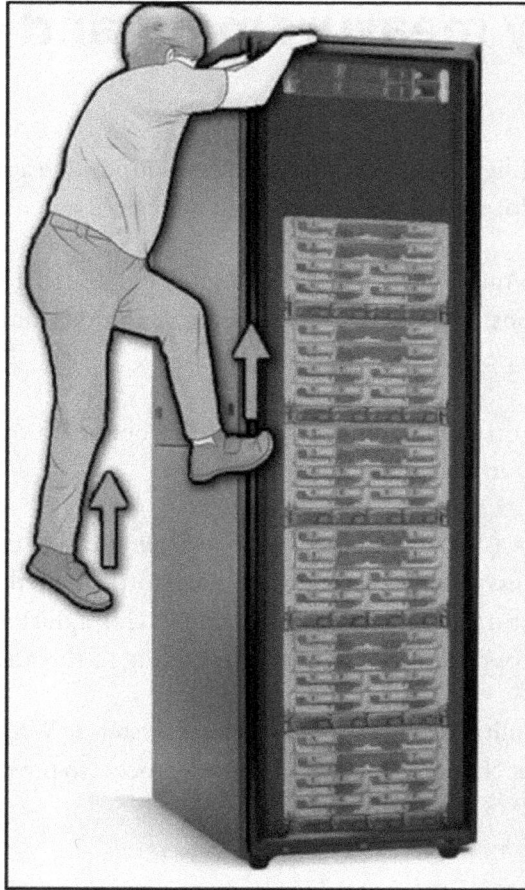

Figure 4-9. A tech is incorrectly trying to achieve IT agility by practicing parkour in the datacenter.

The options were to:

a) Get dedicated servers with HBAs or storage controllers, make sure that the storage system has available ports (which often it did not have room for expansion) and then carve up a LUN for the host, or provide shared direct access via SCSI to the existing filesystem. This would be complex, expensive, risky and have multiple single points of failure.

b) The second option would be to serve up some storage space to external hosts by using hosts that were already attached and enabling NFS or SMB protocols on them to server the data.

Both options lacked either speed in provisioning or isolation and security. So, as you see, the infrastructure that provided scalability and redundancy in a single use case could not easily expand beyond that without caveats and constraints.

The competitive landscape in business is a driver that requires the ability to be dynamic. IT is meant to support this adaptability and the organizations that do it best are more successful than their competitors.

To have IT agility, an organization needs to have an IT strategy that:

1. Adapts to the business' changing needs without requiring a capital project and ramp-up time for every initiative.
2. Has a self-service environment, where the departments within a business can change their internal processes and get the resources they need from IT with no manual intervention.
3. Diversifies the technologies that are used to achieve a business requirement, thereby avoiding lock-in and expanding potential capabilities.

So, this is all good in theory, but how do you use it in practice and how can you measure it within an organization?

Here are four steps that are used to measure the agility of IT in an organization:

1. Define the parameters of measurement.

Some examples are:

a) What is the total turn-around time for provisioning new systems from the initial request?
b) How long does it take to upgrade or change critical systems?

In many situations, the administrative delay in provisioning is orders of magnitude longer than the technical aspects of the task itself. Therefore, making processes more efficient can increase IT agility.

2. Define the factors for measurement.

a) Time
b) Cost
c) Quality

This goes back to the fast/cheap/good discussion we had in section 4.3. The goal in this instance is to reduce time, reduce cost and increase quality. Time and cost are the inputs and quality is the output. However, these factors are not fixed. Rather, they are measured based on the infrastructure. By modifying design and strategy, you can change the measured values for each of these factors.

If you cannot change the design or strategy, but want different values, then you can put emphasis on the other factors.

If you wanted to reduce time for a certain solution, then you could:

a) Hire more people to assist in completing it. This increases the cost.
b) Cut corners and reduce the quality of the output.

3. Establish a baseline.

Document the measured parameters and evaluate them against every add, move and change. Do this for the course of 3-6 months. This average will be your baseline. When doing comparisons to see the change in performance, it is important to compare apples to apples. Create categories for the types of events (i.e.: server provisioning, storage provisioning, application deployments, root cause analysis, etc.).

4. Track the variances and changes.

By seeing the hard numbers of how quickly IT can adapt, provision and respond, you can use that data for strategy discussions, goal setting and long-term planning. It is good to plot changes over the course of every 6 months. One method of doing this is to have a change management and ticketing system in place. Each phase of the ticket can be followed and graphed over time to see how long each action took to complete. Is the bottleneck administrative? Is it procurement? Or is there a technical deficit that could be addressed by training? Look at the personnel resources assigned to the task. How many total hours were associated with the initiative? What is the cost per resource or the net cost for the man-hours of the project? What was the hardware / software cost? After each initiative is complete, is a post-mortem meeting held? What is the project summary and were the lessons learned captured? What was the quality of the output? How can it be improved?

To have agility, there are multiple strategies that must be used. There is the long-term IT strategy which includes a vision, direction and process. Then there is the "rolling plan" that changes all the time to adapt to the needs at hand. The difficulty will be when these strategies are juxtaposed and in vastly different directions. The coexistence of these two strategies is important, even if at times it creates contradiction.

Some risks with agility are:

1. Divergence from long-term strategy while adapting to change in business. This can create a mish-mash of initiatives that are not cohesive. The management of this can then be very

difficult because of the use of multiple technologies that might have overlap in utility, but not in practice, because of planned silos.

2. Reclamation of resources from projects will need to be planned for. Without a planned lifecycle for projects, there will be infrastructure sprawl.

3. By having rolling strategies, the cost to execute many diverging projects can be large and the economies of scale from a long-term strategy and cohesion are lost.

4.12 Chapter Summary

1. A person's personality type plays a large part in how they approach and respond to risk.
2. It is important to understand motivations and interpersonal dynamics when factoring the human element of a design. Incompetence and apathy can be just as dangerous as a malicious actor when access to critical systems is involved.
3. By understanding the available resources and constraints within an organization, you can reallocate areas of focus to better adapt to change. If you assign values of 1-10 to fast, cheap and good, you can get an idea of the viability and variances that an initiative can tolerate. A combined value of 15/30 or more is viable, whereas 30/30 is the "unicorn zone".
4. Understanding the similarities and differences between departmental and regional cultures will help organizational cohesion. This is crucial when conducting mergers and acquisitions. Do not make dramatic changes to culture during a merger, especially if it might spook teams that are prime revenue generators.
5. Cloud services are changing the way IT operates. It is important to plan for and address technical deficit in this area. Do not expect traditional IT to simply "pick it up".
6. By understanding the capabilities, weaknesses, and risks associated with technology that is end of life, or decommissioned, we can effectively repurpose it for secondary utility with confidence. This increases functional capabilities and balances cost and risk to achieve the maximum value from technology resources.
7. DevOps is a modern iterative approach to IT. The most accepted adoption method is the "Three Ways Principle". This consists of: The first way "Systems Thinking", which aligns IT to business processes and promotes efficiency. The second way, "Amplify Feedback Loops", builds communication channels and peer review. The third way is "A culture of continual experimentation and learning" which pushes technical teams to expand their horizons, think out of the box and become masters of their discipline.
8. The desktop computing model is changing; with more services being delivered via SaaS, browser based services and VDI. Migrating to free and open-source technologies for the endpoints has never been easier. Many big organizations are moving in that direction to reduce cost and increase both manageability and reliability.
9. By treating the infrastructure as code, it is easier to scale and incrementally improve over time. Tools like vRA, Chef, Puppet, Ansible and Saltstack can assist with this.
10. Unwritten knowledge from technical staff in the field is extremely valuable and needs to be captured. Start this by working on one process at a time and understanding all the parts of the workflow, optimize it and then go to the next one. Conduct peer review to get input on process efficiencies.

11. As businesses evolve and need to respond to market changes; IT can sometimes become a bottleneck. It is important to have key performance indicators for an organization's ability to execute on IT initiatives that support the business units during these periods of rapid adaptation and response. These metrics can be used to create an improvement plan.

4.13 Chapter Review Questions

1. An organization has recently experienced some critical failures and has recovered, but with heavy losses. The existing CTO has stepped down and the board of directors wants to be very cautious and conservative at this point. They are looking at replacing the role with someone that has an ESTJ personality type. How may this personality type affect the alignment of risk appetite between the CTO and the board?

2. An organization has an ongoing project to implement a new line of business application. They are in the development phase, but the coders are getting frustrated and leaving because they do not have a proper test environment to work in. They have adequate budget for this project and it has been rated 10/10 thus far because no resources have been used. The infrastructure for the test environment is currently rated at 3/10 and no quality output is occurring. The developers have been leaving, so the speed is rated at 5/10.

a) Using the good/fast/cheap Venn diagram and 30-point rating system, how can the project be saved by reallocating resources?

b) Are there enough resources to meet minimum viability?

3. When implementing a DevOps initiative, who would you involve in the feedback process and how would you facilitate the communication?

CHAPTER 5

Guerilla IT for the SMB

"Small business isn't for the faint of heart. It's for the brave, the
patient and the persistent. It's for the overcomer."
--Anonymous (Successful Small Business Owner)

---

*In Canada and the US, 40-50% of the workforce is made up from small businesses. Only half
of all small businesses have a lifespan of more than 5 years. This chapter deals with the risks
and efficiencies related to IT that should be discussed within micro and small businesses.*

5.1 The Risk of Attempting to Operate like an Enterprise

Running a small business is different than running an enterprise. In an SMB, people will wear many hats. The CEO may also be the COO and Director of Sales. The CTO may also be the Director of Security and Chief Architect. Building a business from the ground up needs people that are flexible, adaptive, innovative and work well under pressure. Every deal and every customer is crucial. The revenue goes back into building the business and the operating budget is as slim as can be so that money goes into growth.

IT can be a money sink. It can suck the spirit and life from an SMB if they are not careful. Running IT in an SMB requires a fine line between being frugal and being *cheap*. Being frugal optimizes intelligently, whilst being cheap puts an organization at risk.

Figure 5.1- Some companies cut costs by pirating software. This is a bad idea.

Cheap is: when you get software from torrent sites or back alley channels

Frugal is: not paying for 24/7 software support because the knowledge is in-house and you can pay for support ad-hoc when needed.

Cheap is: having all your clients' critical servers residing in your office "colocation" with residential cable Internet service for WAN connectivity and a couple of old UPS devices for power outages. I've seen this. It's not pretty.

Frugal is: having colocation space that you sublet a portion of to a partnering company to offset operational costs.

When looking at a budget, there are the "must have" and "nice to have" components. This association of priority is a source of risk if not calculated properly. It needs to be based on not only what initiatives help the bottom line, but also what helps reduce the risk of financial loss due to failure, a security event, lack of efficiency or competitiveness. Let's look at some of the key areas of IT spending when looking at an enterprise and then look at options for an SMB.

1. IT Acumen

In an enterprise, there are teams dedicated to specific technologies. There is the storage team, network team, application teams, security team, etc.

In an SMB, there are normally two types of IT staff: generalists that may have some additional knowledge in key areas, or specialists that operate within a strict scope that do not have a wide breadth of experience or exposure to other silos.

The cost of IT acumen that will cover the entire spectrum at an expert level is high. If generalists do everything, then I guarantee that things will not be done in the most efficient way, or the most cost effective. What you will get is a solution that works. Generalists are the duct tape of the IT industry; they can build anything, but it won't always be the prettiest solution. There are many organizations built by generalists that are running purely on luck, duct-tape and python. SMBs run into issues when the generalist duct-tape mentality starts becoming the standard and things like process, change control, documentation and strategy get tucked away a little too deep in the closet.

Specialists and subject matter experts know how to work with specific technologies and apply efficiencies, best practices and standards. However, if you want to get initiatives done that require multiple technologies, then you need an SME for each, or a combination of generalist and SME, or enough time for the SME to learn the additional technologies that need to be supported. The scale and scope of the initiative needs to be weighed so that the right resources can be assigned.

It is much more expensive to pay staff to learn a specific skill, than it is to get a contracted or outsourced SME. That is, unless the skill will be used on a continual basis within the organization, then it makes a lot more sense to cultivate the skills internally. On the flip side, what if the generalist needs to do things multiple times while they learn to get it right? Will they be testing their newfound knowledge in production? Will the business be the lab? This creates added risk and expense. It's good to know when it makes sense to bring in outside resources, as it should minimize risk and unforeseen expenses.

2. Patch and upgrade expenses

Providing applications to users is a huge continual expense. Do you use off-the-shelf software and pay for support and renewals on a yearly basis? Or do you use open-source software and have knowledgeable internal staff that are active in the community for that software, so they can leverage their peers for support? A third option is that you have developers in-house that can write and support the software more effectively than either of the former two options. Let's look at all three options.

a) Off-the-Shelf Software

It is simple to understand the deployment, management and support required to maintain these applications. The software vendor provides guides, reference architectures, security patches and support. Also, there is often peer support in the community. The problem is that things can get expensive for a small business as it grows or needs to pare back operational expenses. What is often done in this scenario is to drop support costs. The organization that does this will immediately open themselves up to potential cyber-attacks using known vulnerabilities. The more popular the software, the more likely it is the case. In addition, organizations must keep up to the latest version and patch level to reduce the risk of cyber-attack.

b) Open Source Software

Open source software has the benefit of having the code vetted by the community that built it. The frequency of patches and updates depends on the popularity of the project. Some projects are where all the cool kids hang out, whereas others are like a bus stop in the desert. People don't hang out there very long, because it's not as sexy as something like the big data processing engine Apache Spark, or the media center project Kodi.

In 2014 when the OpenSSL Heartbleed vulnerability was publicized, there was only one full time contributor for OpenSSL and the OpenSSL Foundation had only received about $2000 a year in donations to help fund the project.

This brought to light the glaring security issue that underscored the entire foundation of secure communications across the world, which is built on OpenSSL. It was not the first time something like that had happened (nor will it be the last), however, it was the most publicized and the vulnerability was even branded with its own logo, (which was the first security threat in history to have that honor).

Figure 5.2- A deserted bus stop with the OpenSSL heart bleed vulnerability logo on the side

Operational management and support of popular and common open source projects is relatively easy with a small bit of technical savviness. However, the more esoteric the application, the more technical skill is required to support it as there may be little to no documentation or community to garner support from. An in-house specialist will also be required to support the application, which may be beyond the capabilities of a generalist. Sometimes projects just halt, due to in-fighting, lack of interest or funding. The cost comes from the time invested in learning the nooks and crannies of the application and its ecosystem.

c)   Custom Built In-House Applications

There are times when this use case makes sense and other times when it does not. If your organization has a development team dedicated to continual support and feature improvement of the application, then it makes sense. If the CEO has a cousin who is a developer (or some other similar situation) and can whip up a solution quickly, then I suggest a full stop!

This is a bad idea. I say again. This is a bad idea.

It is cheap to start, but expensive to succeed. The more the app is used, the more features will be required. This cannot be managed and supported by a single person over a long-term period. The developer must also have a vested interest if the application is critical to business. This needs to be their day job and not a side-project.

Soon you will need dev, test, QA and UAT environments. You need operational support, code auditors and bug bounties.

It is not uncommon for internal closed source applications to become open source to get the benefits of all of these points.

3. Hardware refresh

How often does an organization refresh its hardware? How is that even calculated? For small organizations, or ones with a small IT budget, it can go something like the flowchart in figure 5-3.

This is a decision flowchart for hardware maintenance and refresh that is often used in an SMB. The following outcomes are numbered as per the diagram, so you can follow the logic by referencing that:

1. Good. One less thing to worry about.

This is natural selection in IT. If something dies and you don't need it and it doesn't help you make money directly or indirectly, then let it go.

That token ring switch not working properly? Get over it. Save the power for something worthwhile.

Your legacy MS Windows 2000 server (that had the only working copy of the previous ERP software version) is down? Forget about it. The data was migrated 10 years ago. Consider it decommissioned.

2. This is common, but it can get expensive depending on release cycles.

If you buy something just because it's the new version, then you will get the benefits of new features, performance, etc. However, do those benefits *really* help your bottom line? Something to think about.

3. Make sure you change your chicken bones every few months. Good luck!

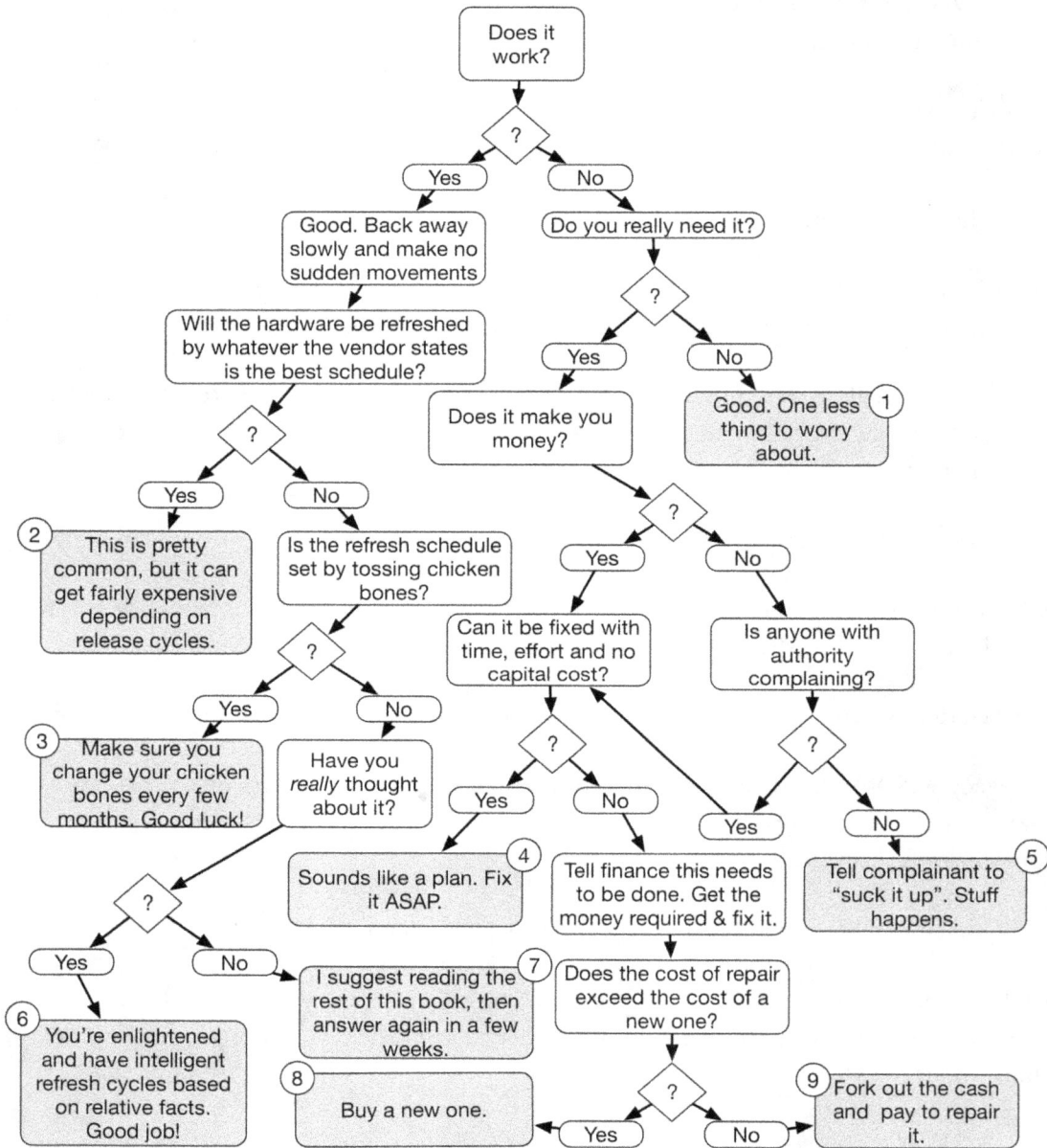

Figure 5. 3 - A decision making flowchart for performing hardware refreshes

This means you have no real strategy and are just guessing at when to refresh your hardware. Maybe it has an inch of dust on it, or maybe the moon is at its perigee with the earth and all the planets aligned, or maybe it's a Tuesday. This will ensure that you're reactive (as opposed to proactive) to any hardware issues and probably ill prepared at that.

4. Sounds like a plan. Fix it ASAP.

In this situation, you have the expertise, resources and time to fix the issue. This is the best state to be in during a failure, as it means that downtime can be minimized.

5. Tell complainant to "suck it up". Stuff happens.

This is the area where someone's pet project dies. The hardware does not contribute directly or indirectly to the bottom line of the company. An example would be an out of warranty videoconference system that is only used between a few offices because it is ultra-high quality. The rest of the company uses a very capable high-quality system that works just fine and will continue to do so.

6. You're enlightened and have intelligent refresh cycles based on relative facts. Good Job.

When this happens, it means that an organization has not only thought about the financial costs over time, but that they have the information to accurately estimate the probability of failure of hardware. This may involve historical stats, industry information, environmental considerations, site specific details, experience, monitoring metrics and dependency risk analysis.

7. I suggest reading the rest of this book, then answer again in a few weeks.

This means that there is no plan, no consideration and no contingency. I do not envy companies in this state, because it means that things will get ugly when they break. If this is your company, this book will help you.

8. Buy a new one.

Something has failed, it can't be repaired and it makes you money. The sooner this decision is made, the faster the issue is resolved and the less of a loss in revenues due to downtime.

9. Fork out the cash and pay to repair it.

This is like number 8, but it is either cheaper, faster, or both, to repair it.

5.2 SaaS and Data Locality

When you are using Software as a Service, or a cloud service provider, there is one very important question you must ask. Where is your data physically located? Does this matter to you? Well it should for a few reasons:

1. If you need to access your data and there is an outage with a non-redundant section somewhere between you and your data, then that is a risk.
2. If your data is not properly secured, then how can you audit access and ensure that only designated people can access the data?
3. If you need to fail-over to another location, how quickly can your data get there?
4. If you decide to leave the service provider, what "exit" penalties must be paid in bandwidth fees?
5. Is your data snapshotted, backed up and replicated? How long will it take to get your business back up and running in the case of a DDoS, regional failure, data corruption or data breach?
6. Is your organization subject to mandatory regulatory compliance that requires data to stay within the country?
7. If your data resides within another country, do the laws in that nation allow for 3rd party access to it without permission or notification?

When the entire architecture stack where your data resides is not under your direct control, then some things may be taken for granted or not thought about until there is an issue. One of these oversights is often the total cost of a full restore. Think of all your data, servers, applications, network connectivity and security. What would be the process to build it back up from a raw data dump? How much time? How many teams? Could it be done on dissimilar hardware or at a different location? What are your organization's fiscal responsibilities to your clients and shareholders? Do you have SLAs that may be breached if outages occur? Have you done comprehensive DR scenario testing?

Many times, organizations are content with DR plans that are provided by a 3rd party, but they do not regularly test these when configuration changes or resource changes occur. This is like driving without a seatbelt because you think good driving, good brakes and airbags will protect you. When your data is not under your full control, then you need to take the extra precautions and plan for the worst scenarios.

5.3 Service Level Objectives on End of Life Hardware

It is important to have leveled expectations of hardware that is past its prime. A change in its role, its workload, environmental factors and the devices it's connected to can affect how much life you get out of it. Where possible, do not use end of life hardware as your primary infrastructure because:

- Security patches may no longer be available and active exploits may be available.
- Risk of failure increases risk of downtime and it may be revenue impacting.
- Redundancy of old hardware can offset failure risk, but it will increase operational costs such as power.

That disclaimer being said, old hardware may be the only thing you have budget for now, or it may be the redundant plan B in case of a failure with primary production systems. Perhaps you want to get some use out of the existing investment to spawn new initiatives that could not have otherwise happened without the equipment.

Some things that you must do are:

1. Create a plan for failure and lifespan objectives.

- Will equipment be immediately recycled when it fails?
- Will you try to repair it?
- What is your cutoff for time and energy on repair and operations before it is non-viable?
- Do you have a cleansing and recycling program or procedures in place?

2. Understand how the hardware will likely fail.

- Based on previous experience and mapped dependencies, how will it affect the system and services for each failure scenario?

3. Create a "spares" inventory.

- If you plan on using EOL hardware for anything of significance, then have spares. Do not expect that you will be able to get more later.
- Standardize your hardware. If you have a mishmash of mixed vendor or custom-built servers, your life will be a living hell trying to figure out what is what and which parts from which machine can be used to repair another. Performance and capabilities will differ, so you won't be able to properly estimate what you can accomplish or where the bottlenecks will be.

4. Have a migration and swap strategy.

- Ensure that the operations team has explicit instructions (and has tested them) on the process for dealing with hardware failures.
- Decouple the roles from the hardware. Use virtualization, containerization and application mobility strategies to ensure that if you inherit newer hardware, you can easily migrate the data and workloads.

5. Understand hidden costs, such as power consumption and operational expertise required.

- What may seem like a good side project with older gear, may end up being more expensive than buying new equipment due to unforeseen operational costs.

5.4 BYOD and Endpoint Management

I'm not going to go into this too deeply, because it is a topic that deserves its own book. However, I will mention it in passing because it is something that should be considered.

If you're a company which employs people, then there is a good chance that they use mobile devices. These could be laptops, tablets, smartphones, smart watches and so on. You may also have standard workstations provided by the company with pre-loaded line of business software on it. It should be understood as to how those devices are used, then managed in a way that works best for all parties involved.

1. Have a Mobile Device Management (MDM) strategy. This includes several of the following considerations:

   - Will there be client VPN connectivity available to access internal resources?
   - Do you care if internal data is left on personal devices?
   - Do you want to know if an endpoint is infected with malware?
   - Do you want to audit or push software packages to the devices?
   - Can your support team access devices remotely if called upon to assist the end users?
   - What apps do staff use to remotely access internal resources?

2. Understand that personal and business use often lack separation in the modern workplace.

   - Users do not want to carry multiple redundant devices just for business (like the work phone, or work laptop).

3. Replacement strategy.

   - If something fails, will you have a temp unit (or several) for users to use right away? If you're using an MDM, then it's simple to push settings and config data out to endpoint devices.

4. Support escalation.

   - How will users be supported in the use of their own devices? Will you take the hard line and say, "Sorry, we only support mobile device vendor A with version x.y.z. of the OS, or vendor B with a.b.c." Or will you say, "We'll give it our best effort, but no guarantees that your old keyboard smartphone will work as expected?"

5. "Shadow IT" gauge.

- This comes down to having the pulse on what a company is organically doing with devices, regardless of any IT policies in place.
- The best way to adapt to a chaotic environment is to understand it and immerse in it. Why are users circumventing policies to get things done? What are they trying to do? Help them explore and get engaged in what they are attempting and maybe they will be more receptive of suggestions from operations that are closer in-line with policies. Or maybe it's an opportunity to change policies to adapt to the users.
- This is not so much about the specifics of MDM and BYOD, but rather that the people who manage IT should have regular communication with the people consuming it. If this communication is not present, then the methods used by users become a division from operations, which is difficult to reconcile and becomes a major risk.

5.5 Organizational IT Acumen

The IT department within SMBs is all-encompassing. Requirements come into it and they must turn those into solutions. There is nothing out of scope. A request may be as small as getting new printer cartridges to something as large as re-architecting the entire IT strategy to capitalize on unproven emergent technologies (that *may* offer a competitive edge in a downturned market).

The solutions are only as viable as the IT staff is capable. How well can your organization deliver on a task? How complex of a task can a single person take on? How well do the staff work together? What are the staff skillsets and what overlapping areas are there? Can a single person perform any task, or are there specific tasks that can only be done by certain people?

This can be thought of much in the same way that applications are associated with hardware. How do you protect against application and service outages on hardware? Decouple the hardware and create a layer of abstraction, like virtualization, then distribute the workload.

Figure 5.4- Who are the knowledge hoarders in an organization?

With people, it's the same process; decouple roles from a specific person so that tasks can be distributed across a group without issue. Keep the knowledge base of the team spread across the team.

A simple test to gauge your current knowledge imbalance is this:

1. Inventory all tasks and skillsets that may be required for day-to-day operations, infrastructure management and special projects.
2. Create a chart to see who in the company has which skillsets.
3. Rate how important each of those skillsets are within the company.
4. Determine the impact that a deficiency of each skillset would have to the company.
5. This will give you a sense of who needs to cross train and who is hoarding knowledge. If you do not know this, then a departure of an individual may force you to re-invent the wheel each time you need to hire a replacement.

Also, make sure that you have training and documentation strategies for technical and professional growth or you may lose your top people as they strive for greater things.

A very important piece of knowledge that is often neglected and not reported, is where the bodies are buried (so to speak). Everyone always knows of a precarious piece of technology or a process that does not work. Provide fail-finder bounties. Reward people for letting you know of risks within your company. The greater the risk identified, the greater the reward should be. Then assign an ID to the risk and track it in the risk register.

5.6 The Dangerous Appeal of Consumer Grade Hardware

It may be tempting to go to Wal-Mart or BestBuy and buy "new" equipment that is consumer grade. However, there are many considerations other than price that need to be accounted for.

1. Security

There are constantly massive bugs and exploits being found in consumer grade hardware. Every vendor has some bugs, but the level of carelessness seems to be exponential in hardware that is marketed as SOHO and SMB. For instance, in 2016 in the United States, the Federal Trade Commission charged ASUSTek Computer Inc. because of their negligence in security for their hardware. They ordered that ASUS would be required to establish and maintain a comprehensive security program subject to independent audits for the next 20 years.

2. Cost of failure

What is the total impact of a failure, or a breach? Take the time to determine what a worst-case scenario looks like. What would it do to the company?

Think of it like this: if you flew in a plane that had doors made of paper, would you be concerned while flying into a hurricane? That is essentially what you are doing with consumer grade hardware. The contents of the airplane are your business, the doors are your security perimeter and the hurricane is the internet. This is not to say that the plane needs to be made of titanium, nor is top-of-the-line enterprise gear required.

If expectations are shifted to realistic levels, then failure can be compensated with planning and redundancy (just as in any environment). However, having realistic expectations can be difficult if the hardware is not researched thoroughly and the limitations and risks are not fully known.

3. Performance

When evaluating performance, you must have specific use cases and workloads for which you are testing for. This is especially true in the SMB market because manufacturing tolerances can be greater and accountability is non-existent. Because of this, you may have performance discrepancies in the exact same models of the same units.

Have a burn-in period and a standardized test regiment to ensure the capabilities match the specs.

After your evaluation and investigation of hardware, you can determine the costs vs. risk of consumer grade and enterprise grade hardware in all aspects of your environment. The fact is that consumer grade hardware is much cheaper in some regards and more expensive in others. Capital costs are only one metric to consider.

5.7 Effort, Cost and Knowledge

To understand the total cost of an infrastructure transformation project, you must factor in several things. First is the business objective. What are you functionally trying to do? Then based on your current infrastructure, where is the gap? How can you get from the existing state, to the desired state?

If you augment your environment, what are the options for hardware? Is there any existing risk that would still be present if an augmentation project occurred?

If you add net-new hardware to replace existing equipment, would it address this risk? Would this process create any new risks?

Can the hardware be allocated from another part of the organization where the capital costs have already been invested? If not, can the equipment be purchased refurbished? Or does it have to be purchased new? Is there an option for trade in credits if the existing equipment is replaced?

For either of these options, how long will it take? How much will it cost, and who will do the legwork for this project? Once the end-state is achieved, will the current operations staff be responsible for day-to-day management? If so, have they received adequate training and/or certification? Taking these factors into consideration will provide an overall plan for each path available. Then the effort, cost and knowledge required to get to the end state can be evaluated for each path.

5.8 To Do This, You Need That

Look at technology from a business solution perspective. Match up solutions and alternative solutions to needs, not wants. It is important to see what other similar organizations use for applications and what their processes are, but understanding your requirements is more important than implementing an identical solution.

Following suit and using the exact same applications as everyone else in the industry is the easy way out, but it is not necessarily the cheapest or most efficient. There may be open source solutions that can provide 80-90 percent of the same solution, or a SaaS offering that reduces the infrastructure and management overhead to nothing. Who knows? Your selection of a different solution may end up being a competitive advantage for your company.

There is no "one answer" to every environment. However, by taking stock of what your business is trying to accomplish, you can evaluate all the possible options and weigh out the risks vs. gains. Sometimes it makes sense to think differently and take a different perspective so that you can see what you're missing, or to justify your decisions.

5.9 Chapter Summary

1. Know your budget and risk tolerance. Use these as your guides and justify your decisions with insight and facts.
2. Always have a firm grasp of your data. Know where it is and how to keep it safe. Have a plan for BC/DR and understand how limits imposed by SLAs that affect your designs.
3. When using old hardware, understand to the best of your ability how and when it will die and what the failure will affect. Avoid it if you can.
4. Get a feel for how users use the devices in your environment. Work with them and not against them to create a common ground. Adapt to what is needed or risk the detriments of shadow IT.
5. Map out the skillsets and capabilities of all IT staff in your organization. Make sure that there are several people that can perform any required task. No lone wolves or knowledge hoarders, as they put your organization at risk.
6. Consumer grade hardware has hidden risks from security vulnerabilities to less stringent manufacturing tolerances.
7. When designing an infrastructure, see what your working resources are in people, technology and budget. Some plans require a gradual process that is multi-phased with dependencies. Each time a phase is achieved, it changes the state of the infrastructure and all interacting systems will need to be analyzed to accurately understand dependencies and risks.
8. There is more than one solution to a problem. Don't be afraid to experiment and research other options beyond the most obvious one.

5.10 Chapter Review Questions

1. Your company has just acquired another organization and you are responsible for creating a cohesive and standardized hardware strategy, but you have no capital budget for new equipment. Your existing company has invested heavily on security, whereas the acquired company has more than plentiful consumer grade hardware. How will you determine what is used and what is not?

2. The company that was acquired has many remote users that need access to certain applications. These applications can be consumed as a SaaS offering provided by a third party. How can you ensure that the business will continue to operate if the service provider has an extended outage?

3. The acquired company has some critical legacy systems that are in a region on the other side of the world and maintained by local staff with intimate knowledge of its operation. Since the acquisition was announced, there have been rumors that the operational staff may soon depart to greener pastures. What would you do to ensure the operational integrity and availability of these legacy systems?

CHAPTER 6

Math and Risk

(or, how I learned to stop worrying and love logic)

"Pure mathematics is, in its way, the poetry of logical ideas."
--Albert Einstein (Nobel Laureate in Physics)

How can you be sure that a plane won't crash on take-off, or a bridge won't collapse under the weight of a traffic jam? By understanding all the known factors involved, the probability, the effects and costs, you can make more informed decisions that affect key design decisions. This chapter looks at the formulas used in various industries and how they can also be used in IT.

6.1 Quantitative and Qualitative Risk Assessments

When you're designing an infrastructure, you must take many factors into consideration. How it is ultimately built depends on what the accepted risks are. You can build multiple models to determine the far-reaching effects of every decision, but how do you know which model is the least wrong? The answer is logic and probability.

If event A happens, then it will cause negative effect Z.

If event B happens then it will cause negative effect Y.

If you want to protect the environment from negative effects, then surely you would do what it takes to mitigate the risk of event A or B from happening. However, you often only have the resources to protect from one of these. Which one will it be?

6.1.1 The Qualitative Approach

Using a qualitative approach, we can say:

In our personal experience, we've never seen event A happening, but we have seen event B happen. Therefore, event B is more likely to occur, so we will protect against that.

The flaw with this approach is that the data set is small and the reliability is in question as it is anecdotal. Anecdotal evidence of statistics can increase in reliability if you have multiple sources that validate it. Quiz 100 people and ask them the same question; have you experienced event A or event B more?

Is there a flaw in this approach? Let's see by making it a bit more real.

A datacenter migration is occurring and two methodologies, each with their own risks are identified:

Methodology 1: Migrate data over a dedicated MPLS link.

The amount of data to be migrated over the wire is extremely large and based on initial estimates, it may take up to a month to complete. The total available time allocated for the migration is exactly one month. If the timeline is not met, then power and network connectivity will be cut and the migration will not complete. This will have drastic consequences for the organization.

The risk is that the data migration will not complete within the required timeline.

The event "A" is that the data transfer will not complete on time.

The negative effect "Z" is that drastic consequences will occur.

Methodology 2: Perform an offline data migration.

An alternative method to migrate the data is to perform an offline dump of the data, then physically move it to the new datacenter and import it. The process for doing this is not well defined and there are many complex dependencies that could be affected. The entire infrastructure would have to be shut down at the source datacenter, then the data would be moved and imported.

The risk is that business will be impacted during the scheduled maintenance window. A secondary risk is that the procedure for performing this maintenance is not well documented and there may be error that occur in the procedure, because it has not been performed by the organization before.

The event "B" is when an unplanned outage occurs.

The negative effect "Y" is that there will be outages, possibly for an extended period, thus affecting core business.

Now ask the staff doing the migration if the negative effect "Z" from event A is more probable to occur than the negative effect "Y" from event B. Then determine which one is more impactful, A or B.

If the probability between event A and event B is the same and so is the impact, then it doesn't matter which methodology you choose.

If the probability between event A and event B is the same and event A has a greater potential impact (from negative effect "Z"), then the choice to follow methodology 2 and protect against negative event "Y" is the least wrong option.

If the probability that negative event "Y" occurring is greater than the probability of negative effect "Z" and the impact is the same, then the least wrong option is to follow methodology 1. The staff doing the migration are basing the probabilities on their own experiences.

Note: The term "least wrong" is used as opposed to the term "right" or "correct". This was a conscious choice to illustrate the fact that the options given are not all the options possible. Also, the possibility of an event occurring or not occurring, is an estimation that is subject to the laws of probability.

What if we expand the sources of probability validation to more people, say 100. Now let's go to the local pub and ask the first hundred people we see, which event is more probable. Is that a valid survey? No, it is not, because the hundred-people selected may not work with the technology directly. However, if the survey respondents worked directly with the technologies in question, then the results are more valid.

6.1.2 The Quantitative Approach

A quantitative approach focusses more on the hard numbers. If there is enough data to use it effectively, then it is the recommended method. Here is an example using the same scenario as before, but with the assumption that both methods have been performed in the past and there is some historical data:

Migrate data to new datacenter via a dedicated inter-site link.

Known data:

- 100Mbps MPLS connection
- Data to be transferred 20TB
- 30 days maximum time available for migration

At 100Mbps, it would take 20 days (approximately 1TB per day) to complete the data transfer.

The slowest data transfer rate that will allow for the completion of the migration within the timeline is 70Mbps, which would complete in 29 days.

If there is an SLA in place from the WAN provider that guarantees a minimum average bandwidth greater that 70Mbps and allows for short bursts beyond 100Mbps, then this improves the reliability of success. However, looking at historical metrics we see the following:

a) Across 100 sites with the same WAN provider and an SLA for a guaranteed average of 80Mbps (100Mbps peak), the average network utilization is 70-90Mbps.
b) There have been 10 outages for the past year across all sites.

c) Outages have been no longer than 4 hours at a time.

d) No locations have experienced more than 1 outage per year.

Based on this information, we can discern the following:

- The performance SLA is consistently being met.
- Over a 1 year period, there is a 10 percent chance of a site outage.
- In a 30-day period, there is a 0.833% chance of a site outage.
- The maximum impact is a 4-hour outage.
- The maximum tolerable combined outage time before the transfer time is insufficient to complete data migration (assuming 80Mbps guaranteed average) is 4 days.

The probability of negative effect "Z" using a scale of 1-10 (1 being the lowest and 10 being the highest) is a combination of the following probabilities:

1/10 that an outage will occur.

1/10 that the impact duration will be more than 4 hours.

1/10 that the total combined impact over the 30 days will cause an average of lower than 70Mbps to occur.

There are other factors that can be added to the list of probabilities, such as the possibility of the application-level replication being throttled to less than wire speed, or the chance that equipment may fail in the process.

The quantitative approach is preferable when you want to ensure the highest confidence for the probability calculated. However, this is not always possible when there is insufficient information or historical statistics. In those cases, it is recommended to have a qualitative approach with multiple data sources.

6.2 The Dempster-Shafer Theory of Belief

I am by no means a math-first person. However, I like ideas that happen to be conveniently described by numbers. Some people use the proverb "the proof is in the pudding" to say that something can only be proven true, if it is observed in the way it is meant to operate. At least, that's my interpretation of it. Explained another way with the pudding example: the proof that a pudding is good is the fact that it gets eaten.

> "The proof is in the pudding" dates to the 14th century when the phrase was originally: "Jt is ywrite that euery thing Hymself sheweth in the tastyng". Also, "pudding" was historically made with boiled meat and blood. This isn't your modern Jell-O pudding. Suddenly, the proverb has darker overtones as you wonder what "proof" is in the pudding. In Shakespeare's tragedy "Titus Andronicus", Chiron and Demetrius are killed, baked into a pudding and served to their mother as an act of revenge for their own evil and terrible deeds. The proof, indeed, is in the pudding.

The "proof" we are after when dealing with risk, needs to be found before the pudding is eaten, or in pudding that has been eaten by someone else first. We do not want to risk eating bad pudding. If we look at it from another perspective, how sure are you that what you believe to be true, is in fact true? What is the likelihood that your facts are false?

This is where the Dempster-Shafer theory (also called evidence theory) comes into play. It is a mathematical theory that allows you to get evidence from multiple sources and combine those into a degree of belief. I will present it in a limited fashion so that the goal and the method of it is represented more than the mathematical formulas. Below are the main components of it.

6.2.1 Components of the Dempster-Shafer Theory

The components that are referenced within the definition of other components, are italicized.

Hypotheses

A series of propositions based on observations.

Frame of Discernment

This is the set of all *hypotheses* being reviewed.

Mass Value

This is the amount of belief that the reviewer has in a *focal element*.

Focal Element

This is the set of hypotheses with a *mass value*, within a *body of evidence*.

Body of Evidence (BoE)

This is several focal elements with mass values.

Plausibility

This is the degree of truth that we believe is within a *focal element*.

6.2.2 Example Use of the Dempster-Shafer Theory

Here is an example of a use of the Dempster-Shafer theory:

When creating a risk scenario, you need to determine the probability that the risk will occur. This can be done quantitatively if there is enough historical data to see the frequency of occurrence over time. If there is not enough data for that, then you need to determine the plausibility that it could occur.

Plausibility is determined by a level of belief plus a level of uncertainty. What is the plausibility that a disgruntled worker will cause damage to an infrastructure after being fired? I will build one risk scenario to evaluate this.

Risk scenario:

A disgruntled worker gets fired on a Friday and over the weekend all the data on the company's servers gets deleted.

What is the probability that the disgruntled worker had a hand in the malicious action? Let's look at the evidence against our hypothesis.

Hypothesis: The disgruntled worker directly contributed to the data loss.

Focal Element 1: The worker was known to act out in anger.

Focal Element 2: The worker had animosity towards the company.

Focal Element 3: The worker had administrative access to the environment remotely.

Focal Element 4: The worker had nothing to gain from performing the digital vandalism.

Each focal element has a mass value (subjectively assigned) associated with the evidence proposition. It also has a negative proposition (like a devil's advocate statement). The combination of the proposition value and negative proposition value must add up to one. If it does not, then the remainder is classified as uncertainty. The belief plus the uncertainty equals the plausibility. See the table below for an example of values based on the 4 focal elements above.

Focal Element Propositions	Mass	Belief	Plausibility	Yes	No
The worker was known to act out in anger	0.5	0.5	0.8	0.8	
The worker was not known to act out in anger	0.2	0.2	0.5		0.5
The work was not known well enough to comment on	0.3	1	1		
The worker had animosity towards the company	0.6	0.6	0.8	0.8	
The worker did not have animosity towards the company	0.2	0.2	0.4		0.4
The worker did not exhibit any feelings either way	0.2	1	1		
The worker had administrative access to the environment remotely	0.2	0.2	0.8	0.8	
The worker didn't have administrative access to the environment remotely	0.2	0.2	0.8		0.8
The level of access the worker had is unknown	0.6	1	1		
The worker had something to gain from performing the digital vandalism	0.1	0.1	0.3	0.3	
The worker had nothing to gain from performing the digital vandalism	0.7	0.7	0.9		0.9
It is not known if the worker had anything to gain from the digital vandalism	0.2	1	1		
Plausibility				2.7	2.6
Number of propositions				4	4
Total plausibility that worker was involved				67.5%	
Total plausibility that worker was not involved					65%

The above table is the body of evidence. The focal elements work in the same way that a detective would work. You look at all the evidence and based on your belief of each item, you combine them and determine how plausible the hypothesis is. Sometimes you have evidence which contradicts the hypothesis and the negative proposition ends up having a higher value. This is the case in the focal element that states that the worker did not have anything to gain based on the malicious act.

The frame of discernment in this case is that there is a 67.5% plausibility that the worker was involved with the malicious action and a 65% plausibility that they weren't. The reason that the two values do not add up to 100%, is because of the degree of uncertainty in the focal elements. The less uncertainty, the closer the values are to adding up to 100%.

There are several methods of combination which consider conflicting information and I acknowledge that the method presented is not the only one, nor the best. It is, however, a simple process to improve decision making and a way of validating beliefs. This, in turn, allows for a higher degree of certainty in risk criticality values.

6.6 Single Loss Expectancy (SLE) of an Asset

The SLE is the financial impact associated to an asset when an event defined in a risk occurs. If, for example, there is a risk that a router will fail, what will the cost in replacement be? If there will only be a portion of the asset that is replaced / repaired, such as the power supply, then that is defined as the exposure factor (EF). The impact of a risk event on an asset is expressed in a decimal value between 0.0 and 1.0. A complete loss and replacement of an asset is an EF of 1.0.

SLE can therefore be expressed by the following means:

SLE = Asset Value (AV) x Exposure Factor (EF)

Using the power supply example in the router, we can say that the router is $5000 and the power supply failure will reduce the value of the device by 30%, then the EF is 0.3.

For the router, with a risk that the power supply will fail, the SLE is:

SLE = 5000 x 0.3

Thus, the SLE for this risk is $1500. This can be used to evaluate the financial impact of risk with all associated assets that are affected by the event and then prioritize risk response strategies accordingly.

6.7 Survivability Analysis

How long does it take for an asset to cease functioning? What is the lifecycle for that asset? Is the lifecycle matched to the technical lifespan of an asset? What if certain factors were changed that increase or decrease the technical lifespan, would that change the lifecycle used for the asset?

These are the questions that can be solved by performing a survivability analysis on assets. The answers are in the statistics. What has happened in the past and what can be expected in the future based on that data set?

The first thing that you need is an inventory of all your assets. Use a change management database (CMDB) for this. If you don't have one, then make one. It's important to have one, regardless of the size of your organization. In fact, the smaller the organization, the more important it is since every dollar must be accounted for and expenses justified to optimize the IT budget calculations.

In the CMDB, track every detail of the asset; purchase date, where it was from, vendor, model, firmware, environment details, location, power consumption under no load and with load, maintenance work, changes, failures, update history, user access, role, supporting workload.

Second, you need a Network Monitoring System (NMS). Monitor all metrics over the lifespan of the asset. Monitor resource utilization (CPU, memory, network, storage), latency, neighboring devices via LLDP / CDP.

Third, you need a Log Management System (LMS). Direct all logs from the asset to the LMS and ensure that it is easily searchable and can provide event correlation.

Now, every asset that is managed will have a default lifecycle based on the type of asset it is. Compute may be 3 years, SAN maybe 5 years, etc. So how do you know if you are getting the best value for your money? Can your compute last 6 years and your SAN 10 years? If so, then you have just doubled your ROI (return on investment) by delaying decommissioning.

The lifecycle is often determined by the vendor, and not the customer. This benefits a company that needs recurring revenue from existing customers, as it is always more difficult to acquire new customers. Vendors have support lifecycles to ensure that older solutions are shelved and new ones are offered that are incrementally better than the last version.

If you want to decouple the vendor lifecycle from the technical lifespan of the solution, then you need to get some very detailed metrics, using the information gathered from the CMDB, NMS and LMS.

There are two common methods that you can use to get data for an asset:

1. Operate the asset in full production until it dies and can no longer perform its function.
2. Operate the asset in full production until the vendor defined lifecycle suggests decommissioning, then operate the asset in a non-essential role until it can no longer perform its function.

> It is important to note that a valid data set for an asset requires that it run until it no longer functions. It must die.

When an asset dies and is decommissioned, you can then review the history of its life. In the 2004 sci-fi movie "The Final Cut", Robin Williams is a "cutter"; he reviews the entire video footage of a person's life and makes a 'greatest hits' video montage for funerals. This is similar to what we are doing with the asset metrics over their lifetime.

Key metrics are plotted:

1. The environment it is in.
2. The workload on it.
3. The failures it encountered over its lifespan.
4. The timing of the failures it experienced.
5. The financial cost associated with each failure.
6. The TCO over its entire lifespan.

The main metrics of interest for this exercise are: the environment, the workload, the cause and the timing of the failures experienced. From these metrics, with a large enough data set, you can determine the probability of a future failure for other assets that have similar properties.

This is done using the Kaplan-Meier estimator, which creates a statistical model of this data. There are several formulas that provide similar results, such as: Greenwoods formula, Cox Proportional Hazards Model and the log rank test. I'm going to claim ignorance on the benefits and drawbacks of each and simply say use a software tool to plot your results.

Some software that you can use to create Kaplan Meier curves are: SAS with the proc lifetest procedure **www.sas.com**, Medcalc **www.medcalc.org**, XLSTAT with either the Biomed or Ecology modules **www.xlstat.com**, or to go with an open-source solution, you can use the lifelines package with Python **lifelines.readthedocs.io**.

The output will be a set of steps that go from 100% survival to 0% over a period of time. The steps are the intervals in which the data is gathered and the percentage is reflective of the number of assets that have been added to the data set. The time interval period could be daily, weekly or monthly, depending on the resolution of information required. The total failure of the asset is recorded and plotted along this path. The more assets in the data set, the more accurate the curve is. If you only have 4 assets in the data set, then the percentage of survival will increment in 25% jumps, whereas 100 assets would allow for a 1% increment. If the time intervals (steps) are too great or too small, then the resulting curve may not represent a useful visual model.

For simplicity sake, you only need 3 values per asset:

a) The group.
b) The time period it is monitored.
c) Whether or not the asset has failed.

If an asset is currently being monitored and it has not failed, or it has been decommissioned, then it is censored from this analysis. See the table below for a sample data set of two groups.

Here we have two groups that, over a period of time, have experienced a total failure. A group is a collection of assets that have a similar property. You can have a group for every property of an asset. Therefore, you can have many more groups than assets.

Two example groups would be environment and workload type.

Group	Time	Total failure
1	300	1
2	500	1
1	250	1
2	325	1
2	275	1
1	100	1
2	400	1
1	250	1

To the left you see a Kaplan-Meier plot with only 4 values in each group data set.

Group 1 is the black line and group 2 is the grey line. These lines show the difference between the lifespan of the asset based on the properties.

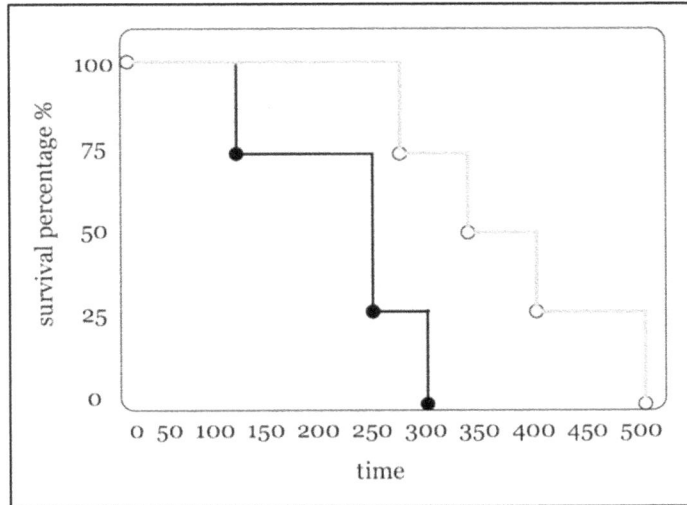

Figure 6. 1 -Survivability Analysis – orthogonal stepped graph

An example comparison would be to see the difference in lifespan of the asset based on environment. For example, Antarctica vs. Bangalore.

The more data points we have, the clearer the picture is on the lifespan of the asset based on each property.

If we smooth out the stepped graph descent, we can see a bit more clearly how the paths are similar and divergent.

In groups 1 and 2, the rate at which the devices fail have a similar curve, but the lifespan is about 200 units longer for group 2 set.

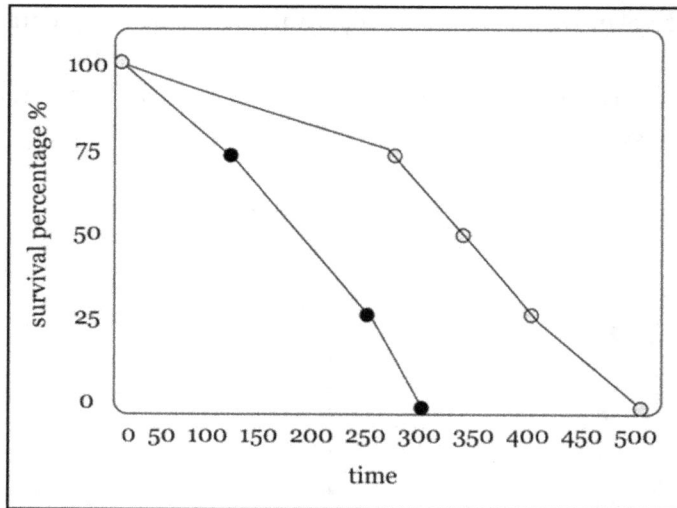

Figure 6. 2 - Survivability Analysis - bezier curve graph

At each time interval, the probability of total failure is the inverse of the survival percentage. For example, looking at group 1, at 250 time units there is a 75% chance of total failure. For group 2 at 250 time units, it is 0% chance of failure. If we use the smoothed graph, it is approximately 20% chance of total failure for group 2 at that time point. Here are some sample groups:

You can evaluate the differences in the failure rates between groups to find some interesting insights between assets that should be the same. For instance, if you have a massive amount of failures in the Sao Paulo office in a short period and you have a similar curve to failures caused by Capybaras then you have a big rodent problem at one site.

By seeing the differences between groups, you can determine things that hinder or help the lifespan of an asset.

Then as new assets get added with specific properties, you can estimate the lifespan of it based on the combined statistics.

Environment name	ID
Des Moines datacenter	1
Bangalore office	2
Sao Paulo office	3
Antarctica field office	4
Shenzhen office	5

Workload	ID
High Performance Computing	6
Bitcoin mining	7
ERP System	8
Cache server	9

Failure Cause	ID
Power Supply	10
Capybaras chewing cables	11
Capacitor leak	12
Exploding battery	13

Figure 6. 3 - Angry capybara

The lifespan of an asset can be calculated with considerably more accuracy by going through this method. This will allow you to do long term planning and decouple the vendor lifecycles from the asset lifespan. Sometimes it will be longer, sometimes it will be shorter, depending on the properties of specific asset.

The difference between the curves of groups can also be used to determine changes that should be applied to improve the lifespan.

This is called the hazard ratio. By identifying the differences between the curves and whether it helps or hinders, you can have a positive or negative coefficient. A positive coefficient means that the curve has a rate that achieves 0% survivability quicker and a negative coefficient means that the asset lives longer.

If you identify the groups that have negative coefficients to a baseline group, then see what they are doing more efficiently to make the asset live longer.

6.9 Inherent and Residual Risk

Inherent risk is what you have if no treatment has been performed to address it.

Residual risk is the amount of risk that remains after a treatment has been performed.

Here are the examples of residual risks for some treatments:

Avoiding the risk: This greatly reduces the risk, but it will not eliminate it. There is still some residual risk.

Remove the risk: This eliminates the risk, but may create a new risk in the process. There is no residual risk for the original risk. There may be inherent risks for new ones created.

Changing the likelihood of a risk: The risk probability has decreased, but there is still a residual risk.

Changing the consequence: The risk impact has decreased, but there is no less probability. There is a residual risk.

Sharing the risk with another party: The risk is reduced, but there is still residual risk.

Taking the risk to pursue an opportunity: This is accepting the risk. Inherent risk = residual risk.

Retaining the risk by informed decision: This is accepting the risk. Inherent risk = residual risk.

Ultimately the action that is taken depends on the total costs associated with it. Will the time, effort and financial cost of the risk treatment justify the reduction in risk that it provides?

If we say that risk criticality = impact x probability, then we need to quantify impact. This is done via using the single loss expectancy method (SLE) from the section 6.6. From the SLE we see that impact = cost. This gives us a value that we will use in our next equation.

Inherent Risk = cost x probability

Once a risk treatment is applied, then the inherent risk is reduced by a value called vulnerability. A vulnerability of 1 means that the inherent risk = the residual risk. A vulnerability of 0.5 means that the residual risk is half of the inherent risk.

Therefore:

Residual Risk = cost x (probability x vulnerability)

Now let's walk through a scenario.

An environment has a single storage system and two controllers. It has several tiers of storage, from high performance flash to near-line SATA. The backups in the environment are performed on the same storage system as the high-performance workloads, but on a slower tier. One controller has failed and a replacement is not available for two weeks.

Risk:

During the period that it will take to get a replacement controller, the storage system has no controller redundancy. This means that any maintenance that occurs will take the entire storage system offline. If the remaining controller fails, then there will be no means of recovering data and all workloads relying on the storage system will be unavailable for several weeks at a minimum.

Based on the lifespan of the controller, there is a 50% probability that it will fail, as highlighted by a previously created survivability analysis.

The estimated loss in revenue per day is approximately $1.5 Million dollars. Two weeks of downtime would cost the organization a minimum of $21 Million dollars, but probably higher because of the poor public perception and the time it would take to get the organization back on track.

The inherent risk is:

Inherent Risk = cost x probability

= 21,000,000 x 0.5

= 10,500,000

Residual Risk = Inherent Risk x vulnerability

Possible risk treatments:

1. **Avoiding the risk**: Move workloads onto other storage systems that have redundant controllers. This has an assumption that there is one available with sufficient space. This sets the vulnerability to 0.1 instead of 0.0 because of the time it takes to perform this. During the data migration period, the risk increases because of the additional load on the system, but it reduces afterwards. The residual risk is 1,050,000. The cost is time from internal staff performing the actions. If an additional storage system needs to be added, then that is a large additional cost that should be calculated in.

2. **Remove the risk source**: Move all operations to another facility and shut down the one with the failed storage system controller. This is an assumption that this is possible. This would require another site to have available resource capacity and have enough bandwidth to support the migration from one datacenter to another. Remote application and VPN access would also be required for staff to continue work while the backend infrastructure is repaired. The vulnerability would be set to 0.5, assuming a 1-week migration time over the network. The risk would initially increase as extra load is put on the system for the migration. The residual risk is 4,750,000. The cost is any bandwidth overage that is used for the transfer of data and the loss of use for the destination storage system and compute resources that are being provided for the workload.

3. **Changing the likelihood of the risk**: Reduce the workload on the storage system. Accelerate the acquisition of the replacement controller. If some of the workload can be reduced by leveraging other systems, delaying heavy operation and only using it for critical purposes, then the risk of incident from a heavy workload is reduced. If the resolution of the replacement controller is expedited, then the exposure period is reduced, thus reducing the cost impact. If the load is reduced enough and the controller replacement is expedited by a week, then we can say that the vulnerability is 0.4. The residual risk is 4,2000,000. The cost is the time invested by staff in the risk treatment and the opportunity cost of not running at full workload levels.

4. **Changing to consequence**: Repurpose another storage system as an emergency backup system in case of failure. This does not change the probability of the risk occurring, but it does reduce the recovery time significantly. A two-week downtime can possibly be a one day downtime with reduced performance. We estimate 2 days to set up the emergency backup system. The estimated loss in revenue per day is approximately $1.5 Million dollars. At 1 day of downtime, and a performance hit of approximately 50%, the vulnerability is 0.6. The residual risk is 6,300,000. The cost is the time required to provision the emergency system and the opportunity cost of not using the emergency backup system for its primary function.

5. **Take the risk to pursue an opportunity**: The risk may highlight deficiencies in the IT budget which stakeholders must address to reduce the risk exposure. This is a risky play

and will not change the vulnerability of the risk, but it may help in the future to secure additional IT budget funding. The residual risk = inherent risk. There is no cost except for the time to execute the plan.

6. **Sharing the risk with another party**: By leveraging vendor SLAs and pushing escalations, you can possibly share some of the financial risk associated with monetary loss from a major outage. This is a difficult process, and it may require legal counsel and furthermore, it will not reduce the risk exposure or possibility. Residual risk = inherence risk. The vulnerability = 0.75, but it will be an uphill battle. The cost is in the planning, communications and legal advice given.

6.10 Chapter Summary

1. Qualitative risk assessments are subjective, but the reliability can be improved by increasing the number of sources and their credibility.
2. Quantitative risk assessments are based on hard numbers and probabilities. The more data that is available, the better. When there is sufficient data, these are more reliable than qualitative risk assessments.
3. The truth in your belief of a hypothesis should be based on as much evidence as possible, until it can be qualified as a fact. The Dempster-Shafer Theory allows you to do this mathematically.
4. Calculating the Single Loss Expectancy (SLE) of assets will allow you to determine the financial impact of a risk scenario.
5. By creating a CMDB and logging metrics and properties of assets over time, you can perform a survivability analysis on them to determine the true lifespan of them and incorporate that information into infrastructure planning.
6. Inherent risk is the amount of risk without performing any risk response. Residual risk is the risk remaining after a risk response has been performed.

6.11 Chapter Review Questions

1. Create a risk scenario for a site failure caused by a fire in the datacenter.
   a) Perform a qualitative risk assessment.
   b) Perform a quantitative risk assessment.
2. Calculate the SLE of all assets affected.
3. If a risk is accepted to pursue an opportunity, what is the residual risk?

CHAPTER 7

The Human Element

"Nothing in the world is so compelling to the emotions as the mind of another human being."
--Margaret Floy Washburn (PhD Psychology)

Every organization is the sum of its people. Understanding the psychology of an individual, is like understanding the state of a component in a system. There is a constant state of flux and equilibrium, change and routine, give and take. Every person has a relationship with themselves, their peers and the organization. This chapter is about understanding how risk relates to the human element.

7.1 Decision Theory

Understanding the decision-making process will help you understand the logic in a decision and weed out fallacies that may be skewing perception. This will, in turn, make you more adept at making well thought out decisions and understanding those of others. When designing a risk considerate infrastructure, the decisions that are made must be logical and well thought out. Emotional responses and quick judgment calls should be avoided as they have a higher likelihood (without adequate experience) to provide non-ideal outcomes.

The process that we use to make decisions is complex. Decisions can be put into three categories: rational, non-rational and irrational.

7.1.1 Rational Decisions

Rational decisions are made by assuming that the outcome should be the best possible for the actor (decision maker). They are measured against the possible outcomes and how closely they align to defined goals.

7.1.2 Non-Rational Decisions

Non-rational decisions are made by intuition and how something feels. They are decisions that are derived from subconscious sorting, evaluating and analysis. They are judgement calls that may be difficult to explain if required to walk someone through the thought process. Some of the best non-rational decision makers are those which have a lot of experience from which they can draw upon.

7.1.3 Irrational Decisions

Irrational decisions are ones that do not seem to align to defined goals. They are not logical and do not seemingly benefit the actor. A decision that is irrational is not necessarily a wrong decision, because the outcome may still be beneficial. However, this is only true when unforeseen events skew the expected outcome, or when the motivations or strategies are misunderstood by others.

An example of this is in the Pink Panther movies, Inspector Clouseau makes a series of bad decisions, has bumbling accidents and against all odds, solves the case. It was external influences that turned his irrational decisions into positive outcomes.

Figure 7. 1 - A bad tattoo is an example of an irrational decision

Another example of this is a chess player who uses a misdirection strategy to focus their opponent on one area of the board, while developing another. From the perspective of the outside observer, some moves seem to be irrational decisions, when in fact they are just not understood.

Most irrational decisions are not obscured plots created by geniuses, but rather just poor decisions. They may have occurred because of lack of information, misconceptions, or preconceived notions. Many irrational decisions are also predictable when observing an actor that frequently makes them.

I knew a guy that would make irrational decisions from Friday to Tuesday every week. At first, he was thought to be wild and fun, but after a while he was making predictably irrational decisions. On more than one occasion, he would show up to the office with wearing the same clothes as the previous day and smelling like he had gone swimming in a distillery. That was predictable irrationality.

7.1.4 Rational Decision-Making Overview

If we look at the logic behind rational decision making, it follows this general course of action:

1. Identify the decision.
2. Determine all courses of action.
3. Construct a payoff table.
4. Choose the decision with the highest payoff.

This seems straight forward and in many cases, it is, at least until you factor in risks and unknowns. Then you have a variable payoff table and its correctness is based on the information you have. You can categorize the decisions into the following types of environments:

1. Decisions with certainty.
2. Decisions with risk.
3. Decisions with uncertainty.

A decision tree consists of two elements; circular nodes (which represent uncertain outcomes) and square nodes (which represent decisions). They are connected with branches and terminated with a triangle representing an outcome.

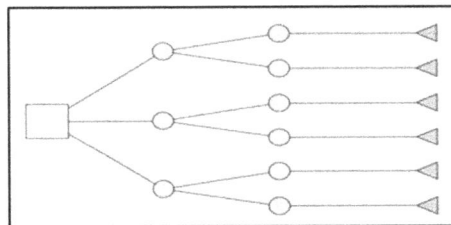

7.1.5 Decisions with Certainty

Decisions with certainty are ones where you know all the possible choices and outcomes. By evaluating the outcomes, you can select the choice that most aligns with your goals. This applies when numbers are static and there is no chance for variation. An example is when you are comparing the price of a specific product between vendors and you want the cheapest one.

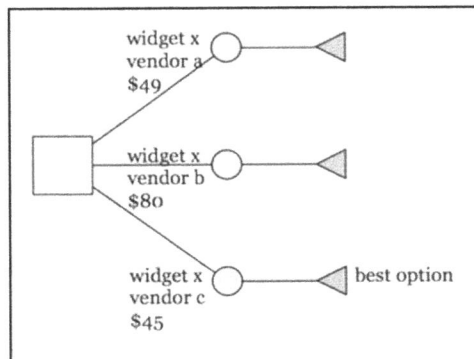

Figure 7.3 - Decisions with certainty

Widget X is to be purchased.

Widget X from vendor A is $49.

Widget X from vendor B is $80.

Widget X from vendor C is $45.

Weighing all the outcomes against the goal of the cheapest one, vendor C is the best choice.

7.1.6 Decisions with Risk

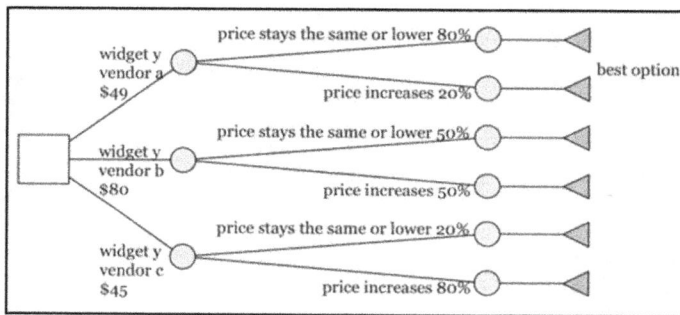

Figure 7. 4 - Decisions with Risk

Decisions with risk are ones where you know the outcomes, but there is a level of probability that needs to be accounted for. An example is when you are comparing the price of a specific product between vendors and the prices may change.

Widget Y is to be purchased.

Widget Y from vendor A is $49 (price stays the same or lower probability of 80%).

Widget Y from vendor B is $80 (price stays the same or lower probability of 50%).

Widget Y from vendor C is $45 (price stays the same or lower probability of 20%).

Weighing all the outcomes against the goal of buying the cheapest one and the probability of the price staying the same, vendor A is the best choice.

7.1.7 Decisions with Uncertainty

When talking about variables that may affect outcomes, there are 'known' unknowns and there are 'unknown' unknowns. Decisions with uncertainty are ones that have 'unknown' unknowns. An example is when you are comparing the price of a specific product between vendors and you don't know if you even need the product because:

a) The client may want to go with an alternative product.
b) The manufacturer will be making the product end of sale shortly.
c) The product cannot be purchased because of a current Vendor of Record arrangement does not allow for it. An alternative product is required to be researched and presented.

There is a big difference between decisions with risks and decisions with unknowns. Scientists have determined that the evaluation of choices posed by the three decision environments use different parts of the brain. The scope of possibilities and considerations for risk and uncertainty are referred to as worlds. A world of risk which is considered a small world, whereas a world of uncertainty is a large world. Most of the decisions that we make in our lives are those of uncertainty. What should I wear today? How will I ensure that my kids have all the opportunities and support they need to succeed in life? How will I stay healthy, fit and maintain a work-life balance while advancing my career and raising a family, writing a book and learning the entire product portfolio for hundreds of different hardware / software vendors while ensuring that time for sleep is allotted here and there… just for example.

The difference between risk and uncertainty is the type of thinking that is required for those decisions.

Decisions with risk require analytical and statistical thinking. Values are used and probabilities are calculated to make the best statistical decision with the known information.

Decisions with uncertainty require heuristic thinking, which cannot be easily evaluated with an algorithm. The best way to approach these decisions are through a methodology or defined process. This way, you do not make a specific choice but rather, you have a set of rules which are continually applied to all choices related to the world of uncertainty. As outcomes develop, you will then can make decisions with risk or with certainty.

When the uncertainty is vast and the time is limited, it is common to switch from a rational decision-making process to a non-rational one. The experience and knowledge of the individual making this decision will dictate whether it is likely a good one. However, the outcome will decide this for sure.

7.2 Framing

A classic example of framing is the glass half full / half empty scenario. Is someone an optimist, or a pessimist? Is something a gain or a loss? There is an old story about a boy that broke his leg. At first his family thought it was a bad thing, but then the military came to town to conscribe all the young men in the village. Now the broken leg is viewed as a good thing, as he no was longer forced to go to war.

Scenarios can be presented in a skewed light to serve an alternate agenda or because not all the facts are presented in the same light. Therefore, it is important to understand scenarios from the point of view of different people.

A network admin will see everything from a network centric perspective: traffic flows, bandwidth, ports, ACLs, etc. A storage admin will see everything from a storage centric perspective: latency, workload characteristics, LUNs, disk pools, IOPs, cache hits, etc. An architect must evaluate a scenario from multiple points of view and understand how they may cause conflicting evaluations and solutions to the same problem.

7.3 Cognitive Bias

Think of a cognitive bias as a break in logic due to holding onto one's beliefs or preferences even after being presented with conflicting information. A cognitive bias will affect the way decisions are made because they often cause one to have a specific view when multiple options are presented. In some cases, this can be a method of reducing the number of choices to fit within a certain belief structure, whereas other times it makes decisions easier, because they are the same every time. Although there are hundreds of documented types of cognitive biases, there are four main categories:

1. Too much information.
2. Not enough meaning.
3. Need to act fast.
4. What should we remember.

I will provide a few examples from each one of the categories.

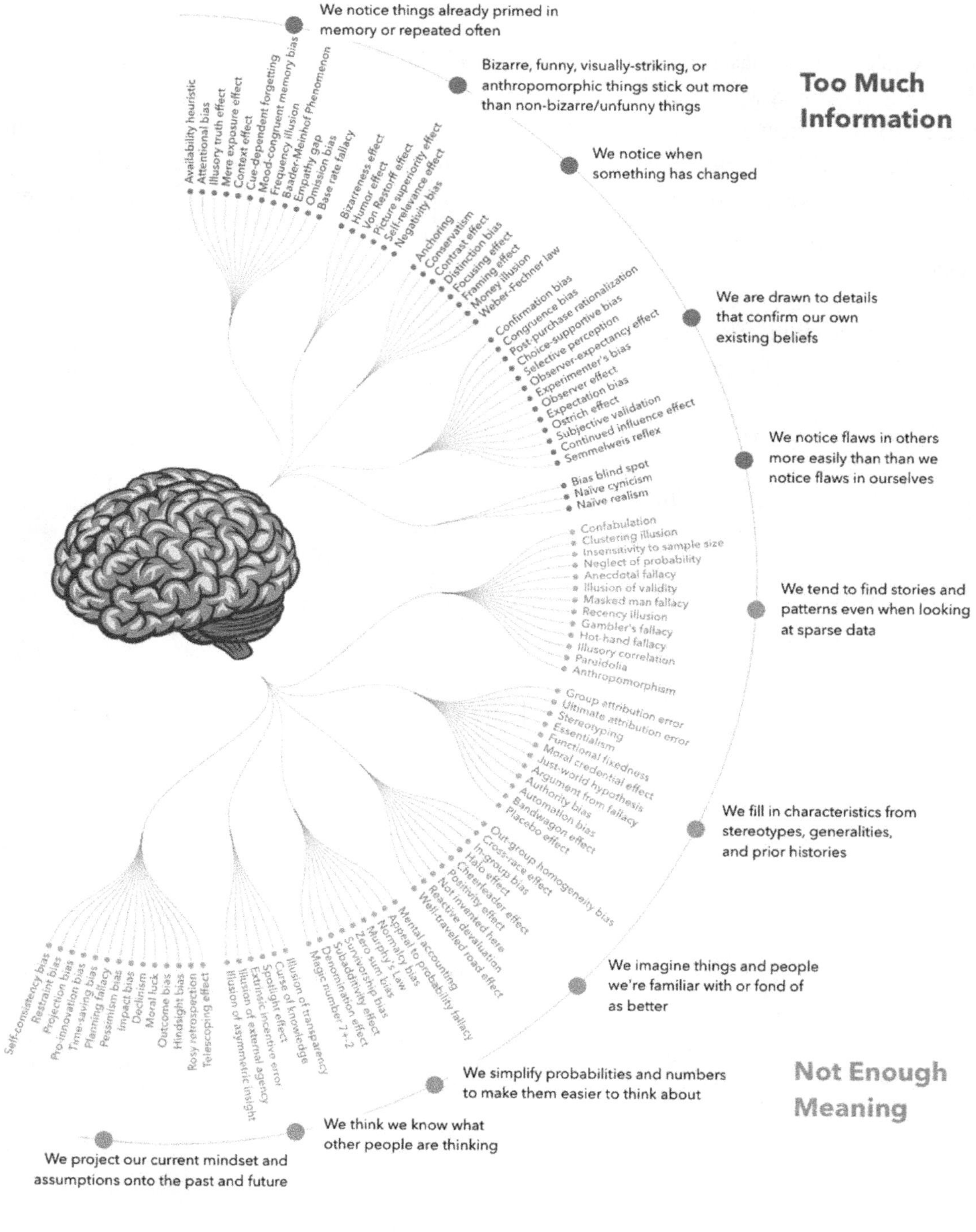

We notice things already primed in memory or repeated often

Bizarre, funny, visually-striking, or anthropomorphic things stick out more than non-bizarre/unfunny things

We notice when something has changed

Too Much Information

We are drawn to details that confirm our own existing beliefs

We notice flaws in others more easily than than we notice flaws in ourselves

We tend to find stories and patterns even when looking at sparse data

We fill in characteristics from stereotypes, generalities, and prior histories

We imagine things and people we're familiar with or fond of as better

Not Enough Meaning

We simplify probabilities and numbers to make them easier to think about

We think we know what other people are thinking

We project our current mindset and assumptions onto the past and future

Figure 7. 5 The cognitive bias codex

7.3.1 Too Much Information

Bizarreness effect: When reviewing a large amount of information, bizarre content is remembered more easily than regular content. An example is when conducting an infrastructure review and you come upon a live token ring network in the ceiling with a running Novell 3.x server that is serving data via an Ethernet bridge.

Anchoring: When a decision is being made and multiple pieces of information are being evaluated, if someone latches onto the first piece of information and puts too much emphasis it, this is called anchoring.

7.3.2 Not Enough Meaning:

Clustering illusion: When reviewing samples of data and there are some small streaks of deviations, someone may overemphasize the importance of those. An example of this is when a root cause analysis of a performance related issue occurs and a brief spike in network utilization is observed, and then stated as the cause. This is also called the phantom smoking gun.

Zero-sum bias: This occurs when someone thinks that for one party to succeed, the other must fail. You see this in the market when vendors start throwing FUD (fear, uncertainty and doubt) around and disparaging their competition instead of highlighting their products or solutions and differentiating themselves from the competition.

7.3.3 Need to Act Fast:

Illusion of control: This occurs when someone believes that they have a greater influence over external events, than they do. An example is when a department has a failure of a piece of infrastructure that is not under warranty. They expect that since it is important to business operations, that it will be replaced. However, unexpected capital expense approvals are not a given. Their feasibility is determined by budget constraints.

Risk compensation / Peltzman effect: This occurs when someone takes greater risks because they have a perception of greater safety. An example of this is when someone is performing a major upgrade of a section of infrastructure and has not taken the proper precautions (with the creation of tested roll-back procedure, etc.) because they have multiple methods of backups in place. They may assume that known procedures can recover in the event of an issue, but they may not have tested them beforehand.

7.3.4 What Should We Remember:

Google effect: This is when someone does not bother learning, or remembering information because they can find it easily via a search engine. This bias causes people to know of things and where to retrieve further information, but not necessarily to understand them.

Modality effect: This bias occurs when what is communicated via speech is recollected more than what is communicated via writing. An example of this is when there is a conference call and what is stated conflicts with what was written via email, but the phone conversation sticks in the mind.

7.4 Social Engineering

Social engineering is a means by which an actor can manipulate a person, or group of people into performing a function that may or may not benefit them. This can be done by choosing a target, gathering information, creating a pretext, contacting the victim and persuading them to perform the function. It is like launching an exploit against a vulnerable mind.

Social engineering is a lot more pervasive than one may think, because it is part of everyday human interaction. When a child wants some candy, they will try to convince their parents to give it to them. When a company wants you to buy their products, they will create marketing campaigns to convince certain demographics to do so. When a counselor or psychologist wants to get their client to open up about a topic, they ask leading questions. When a police or security agency wants a suspect to reveal information, they will use interrogation techniques. When a criminal organization wants to obtain information, knowledge or access that someone has, they will create a rapport with them and convince that target to freely give it up.

So why is social engineering in a book about risk? Because people are the risk, and they need to be considered as such. If an infrastructure design does not consider the massive security vulnerability that people are, then that infrastructure can be easily compromised. Millions of dollars in infrastructure can be bypassed and compromised by a few minutes of conversation with the right people. Social engineering is the most successful method of infiltrating an organization and exfiltrating the data.

Now let's break down some of the techniques and processes.

7.4.1 Gathering information

In the same way that a salesperson gathers information on a potential client, social engineers collect and categorize information from multiple sources. This consists of data mining information from personal websites, whois, and social media accounts of anybody that works for the company.

Examples of types of information are: products and services, locations, job openings, contact numbers, biographies of execs, comments in support forums, etc. Information on individuals could be: email naming conventions, personal websites, their job roles, habits, vacations, where they buy their coffee, etc.

What malicious actors do:

a) The information that social engineers collect will be organized into a database of easily accessible and detailed information.
b) Specially crafted individual emails will be sent to key targets based on the information gathered. It will have links to a malicious payload.
c) Once the payload is clicked on, the computer is compromised.

How to defend against malicious actors:

a) Education. Train staff to know what is suspicious.
b) Employ link scanning in emails.
c) Use stateless desktops that wipe out any system changes that occur, including compromise, upon logout.
d) Segment networks to ensure that servers and support infrastructure are not accessible (except for specific ports for applications) from end user networks.

7.4.2 Creating a pretext

A pretext is the background story. There are threads that weave a story together to make it indistinguishable from truth to the intended target. In incorporates

Figure 7. 6 William Thompson - The first con man

context, understanding of the targets personality, their wants and needs, and then uses those to manufacture a reality that they will accept. Think of what you do every day, with whom you interact, and what those interactions look like. If someone says hello, you say hello back. If you fly on a plane, you show your boarding pass to get on. If the VP of your company asks for an update on a known ongoing critical issue, that gets provided. If a vendor provides an invoice for hardware that was known to be ordered, that gets paid. If a co-worker normally shares files via dropbox to you, then chances are you will click on the link to open it.

The pretext can be woven together from expected social interactions seamlessly. A well-crafted pretext will never get discovered until it's too late, if at all. In the 1840's, a man by the name of William Thompson became known as the original con-man, or "confidence-man". He would dress like an aristocrat and approach upper class citizens in New York pretending to know them, striking a lively and engaging conversation. After a while he would ask them, "Do you have confidence in me to trust me with your watch until tomorrow?" Since he had created a rapport with them, and they were convinced that they knew each other, these unwitting victims would often hand over the watch without question. William and the watch would then disappear, never to be seen again.

Some victims are so convinced by con artists that they cannot believe they have been scammed and sometimes they even defend the scammers after the evidence to their guilt is irrefutable. Other victims are embarrassed when they realize they were conned and don't report the crime to authorities. Companies that are scammed don't always report it because acknowledgement of a crime against the company may affect stock value or lower public opinion. In that case, they would take a double loss and thus experience greater harm than just taking their lumps and moving on. It is because of this lack of reporting that criminal social engineers can continue operations without so as much as a fear of recourse or retribution.

What is done by malicious actors:

a) Communications are intercepted, interpreted and usurped.
b) A dialog is created that is normal within an accepted scenario.
c) Trust is achieved and the victim's confidence is gained, thus lowering their guard.

How to defend against malicious actors:

a) Be present. Do not multi-task or "active-task switch" frequently. Focus on individual tasks in a serial manner, when possible.
b) Remember that if something is too good to be true, it probably is.
c) Keep a watch out for inconsistencies in normal communications, such as changes in the tone, writing style or subtext in emails from known contacts.
d) From a technology standpoint, this is difficult to defend against and requires complex heuristics such as: monitoring communication source locations and routes, medium, frequency, context, message length and writing style.

7.4.3 Other techniques used by social engineers to gain trust.

a) Appeal to the ego.

Praise, showing appreciation, showing respect, pandering. These are all methods of appealing to someone's ego and thus disarming them.

b) Showing a shared interest.

Many people are socially isolated. By sharing an interest (truly or falsely) with a target, the target allows the social engineer into their inner circle of trust.

c) Make a false statement.

People tend to want to correct falsehoods when presented with them. This is a baiting method for establishing dialog.

d) Volunteer information.

By presenting some knowledge that is deemed as privileged, the victim is more likely to act in kind.

e) Assuming knowledge.

By alluding to the fact that certain knowledge is already known by the social engineer, the target is less likely to guard any related information.

Most techniques used by social engineers cannot be stopped by technology. Internal security policies, education and risk governance will reduce the organizational attack surface. Ultimately, the best defense against social engineering is awareness of yourself and knowledge of the areas in which you are vulnerable.

7.5 The Disgruntled Employee

An employee who has unmet expectations or becomes at odds with peers or management, may become a disgruntled employee. Multiple surveys performed by the FBI, CERT and many other security organizations have indicated that based on attacks to an organization, anywhere from 50%-75% can be attributed to inside sources.

Employees that have access to internal systems, have opportunity to damage, delete, corrupt or exfiltrate data. This is because these are requirements of day-to-day operations; users need read/write access to data and the ability to transmit documents to peers, partners and clients.

Here are some techniques, technologies and processes that you can use to protect the organization from the malicious actions of disgruntled employees.

a) Create a clearly defined organizational security policy that outlines the legal ramifications of malicious or illegal actions in the infrastructure.
b) Provide training and awareness to staff of insider threats and the costs to the organization.
c) Monitor the activity of new hires and create a baseline for behavioral analytics.
d) Address concerns that staff have before they escalate to the point of serious conflict.
e) Monitor social media for negative statements about the organization and see if any posts or remarks are attributed to existing employees
f) Inventory and actively monitor all assets in the environment.
g) Have strong password policies in place.
h) Monitor user logins, access and usage. Have alerts setup for non-standard behavior such as after-hours logins.
i) Implement a more detailed level of monitoring for administrative user accounts.
j) Implement a SIEM and aggregate logs to perform event correlation of actions to users.
k) Implement a change management system.
l) Implement a configuration management system and ensure real-time monitoring.
m) Implement user account access monitoring for every access method and endpoint.
n) Ensure a detailed backup and recovery process is in place with offsite backup capabilities.
o) Implement a DLP (data loss prevention) solution.
p) Have a detailed termination process in place for employees.

When employees go past the state of being able to have their issues addressed, no resolution will be able to repair their relationship with the employer. The relationship is tainted and they become disgruntled. They may become withdrawn or try to avoid expressing their level of dissatisfaction. At this point, the employee will consciously or unconsciously begin acts of sabotage such as adversarial conflict, poor work ethic and berating the company. The longer that the employee remains within the organization, the more infectious the negative morale will be with others that work with that employee.

When disgruntled employees start airing their grievances publicly, this opens a huge door for social engineers to provide an ear to listen. By engaging the employee, social engineers can nurture and exploit that relationship which may lead to the employee being used for criminal acts against the organization, data exfiltration, digital vandalism or implantation of logic bombs.

7.6 Investing in Culture

Culture is what makes people work well together. It is the dynamic that allows for interaction between peers, departments, management, clients and partners. It is what gets people up in the morning and excited to go to work. Or, it makes them dread having to spend one more day doing something they hate, with people they may resent.

A vibrant culture does not just pop out of thin air, but rather it is cultivated over many iterations and requires a feedback loop with people in the organization. Questions must be posed, such as: "Are you having fun?" or "Would you recommend this organization to others? Why or why not?" or "What can be done to make this place better?" Culture changes over time and reflects not just the organizational values and priorities, but those of the staff as well.

Some companies that are looking to hire innovative and bright employees are finding that motivation and passion are the keystones to successful hires. They also find that money is not the only carrot on the stick that works to achieve this. In many cases, it is the less tangible factors that win over top talent such as; work location flexibility which allows people to work from the office or home, results oriented work which allows people to specify their own hours and regular social engagements that allow staff to connect with each other beyond the framework of the work dynamic.

People work better with those that they know and can relate to. A stronger social dynamic through trust and understanding will create a more efficient, effective and resilient team. A feeling of belonging and interconnectedness will lessen isolationism, break down silos and create more collaborative opportunities within the organization.

When an organization invests in its culture costs come down, communication goes up and performance goes up. It is a method of protecting against the risk of losing top talent, the risk of having disgruntled employees and the risk of not meeting the requirements of a changing work force.

7.7 When Acquisitions Go Wrong

Companies get acquired, merged and consolidated all the time. A little-known fact is that many of these are terrible failures end up causing massive losses and write downs for the parent organization. What causes something that looks good on paper to fail so miserably in the real world? Let's look at a few possible factors that may cause this.

7.7.1 Culture Clashes

As previously mentioned in the chapter, culture is something that makes an organization successful. Change the corporate culture and now you have changed the recipe for success. A simple way of explaining this concept is the following example:

Chocolate frosting on cake is tasty. Lasagna is tasty. So why not mix two things that are tasty and have more tasty food? Unfortunately, it doesn't work like that. Things that work well separately will not necessarily mix together well.

The ingredients that went into making a chocolate frosting are not the same ingredients that went into lasagna, nor are they complimentary. Separately, they are both still good, even near each other, such as dinner and dessert, but not mixed together.

Culture is a big part of the recipe within an organization. The way that people work is core to their output. A change of culture will change the output, often for the worst. Some people are not very adaptable to change and prefer to remain doing things in the same way at a different organization, rather than changing the way they do things at the existing one, for instance during a merger or acquisition. If the cultures are similar enough, then the merger can create a new similar culture. If the cultures mesh in a positive manner, then you have what is referred to as a synergy.

7.7.2 Buying a Competitor or Other Company and Synergies

What are the overlaps? Which organization does certain things better? How will the two companies provide a greater value to the customer by integrating with each other?

When you have a lack of synergy, then you will see the following:

a) Resentment of management decisions and perhaps of core organizational values.
b) Overlapping roles and departments. If you have multiple departments doing the same thing, but in different manners, then you have a lack of consistency and a loss of efficiency.

7.7.3 The Oracle Effect

Oracle has a history of aggressively buying companies, integrating only parts of them into other offerings, reallocating talent, or dismantling the business.

This heavy-handed strategy is not about creating synergies, but rather of takeover and dominance. Some examples of this are the acquisitions of Sun Microsystems, and Virtual Iron, to name a few.

When large companies are engaged in the process of an of acquisition, the staff of the smaller company may feel some apprehension. Sometimes negative events will occur if the concern is not addressed head-on.

1. There will be an exodus of staff that want to take the uncertainty out of their future. They want to have control over their employment and do not want to chance the possibility of a layoff.
2. Some people thrive in chaos and will use the transition period to push the limits of what is acceptable. Lack of managerial oversight complicates this, especially when similar teams are amalgamated between organizations, but have different roles and responsibilities.
3. Some staff may take the opportunity to secure their positions by allowing certain things to fail and then save it during the crisis. This makes them look less expendable, except to those around them in the know.
4. Some people will do everything they can to slip under the radar and still get a paycheck with the minimum amount of work possible. If there is no possibility of advancement, recognition, or job security, then the incentive to work hard is greatly reduced.

7.7.4 What can be done to increase the chances for a successful merger or acquisition?

The company instigating the merger or acquisition can do the following actions to help achieve a more successful, less stressful outcome for all:

a) Learn about the company that is to be acquired. What makes it great? What is the social dynamic? Who are the movers and shakers? Who are the core people in each department? What software do they use for operations? Who is their client base? Who are their biggest customers?

b) Accept that some staff may not want to join your company, but do not want to leave their role. They probably will be hyper-alert to everything that happens and very critical of the acquiring company. Be cognizant of how the larger company represents itself to the smaller company. An example of this:

A larger company acquired a smaller company that was experiencing some customer attrition due to market changes. The smaller company had to cost cut for years and those who stayed were there because of a sense pride in keeping the company going. During initial meetings between management and staff, the larger company bragged that they "had so much money that they could roll around their money for a week and not touch the same bill twice." Comments like those will almost always be ill-received by staff. The contempt for the acquiring company increased when there was no additional money invested into the smaller company's operational budget or salaries. This acted as a catalyst for an exodus of many top performers within the organization.

c) Work together with staff from both organizations to create the "new normal". Determine concerns by talking to people and address them early, head-on and review them often to make sure everyone is in the know of where things stand.

d) Create engagements between the organizations to share the culture with each other. The first 30 days are the most important to get dialog and interaction started. The second 30 days are also important to start on endeavors that require joint efforts and create new systems and workflows.

e) There will always be attrition during an acquisition or merger. Prepare to combat this with hiring new staff to replace deficiencies in the new combined organization.

f) Continually gage the feeling of the organization. Check-in with old staff and new staff to get an idea of what the feel is and address any issues or concerns that come up.

g) Create organization ambassadors. Have dedicated people on both sides of the merger or acquisition that have the role of engaging people in the other organization and connecting individuals on both sides on a personal and professional level.

7.8 Incompetence vs. Carelessness

There are few things that can cause more damage in an organization than someone with administrative access and a lack of self-governance. Or in other terms, a cowboy that shoots from the hip, or a loose cannon.

Let's look at the two traits that have a similar negative effect within an organization, incompetence and carelessness.

7.8.1 Incompetence Defined

Incompetence is when someone is not capable of performing a job role in its required capacity. This may lead to poor decisions that have massive implications to the organization. An example of this is when someone is tasked to perform a backup of some data, but instead they delete all the data and don't realize it, leaving someone else to discover and recover it if possible.

7.8.2 Unconscious Incompetence

This is when someone does not know what they don't know. If they are so far removed from knowing how the things they are doing work or how to find out that they are doing things right or wrong. Think of doctors before the time of understanding what bacteria or microbes were. Hands would not be washed before surgeries and equipment would not be sterilized.

7.8.3 Conscious Incompetence

This is when someone knows that they lack certain knowledge. Think of an entry level tech support person, being tasked with the coding an update for a complex enterprise application built in-house. This is what happens when someone is given a task that they know they cannot perform without additional resources or specific training to complete the task.

7.8.4 Carelessness Defined

This is when a task or action is performed with a lack of care. Careless acts can occur from people with the prerequisite knowledge for the task, or without it. There are several reasons that a person may be careless. Below are some examples:

If someone does not have a specific methodology defined for performing an act, then some aspects may be missed and cause negative consequences.

If someone is lazy and does not check their work, then there may be errors, inconsistencies or omissions.

If someone performs an activity, but does a poor job at it and cuts corners, then it can be careless or negligent.

a) If someone is distracted, then they cannot dedicate their concentration to a specific task. An example of this is texting and driving. It is impossible to communicate with proper sentence structure and grammar (or even emojis) via text if you are distracted with driving.

Ways of combating both incompetence and carelessness are through increased communication and training. If there is a support structure in place that allows someone to ask a question without judgment or recourse, then it is more likely that staff will make use of it.

Although if there is a great disparity between competence and job role, you may get some interesting requests for information. For example, if an employee asks their manager, "Where are the fire extinguishers located and how quickly can vacation leave be authorized?" then there may be a training opportunity there.

Once gaps are identified by management, peers, or the individual employee, then training is the best route to overcome incompetence and to enforce accountability thus reducing carelessness. If the costs of training are ever an issue, then calculate the costs and fallout of an event or issue caused by the lack of it. If the potential for loss is greater than the costs for training and the probability is beyond an acceptable threshold, then it makes business sense from a risk mitigation perspective, to ensure that it is provided.

7.9 Employee Workload as a Risk

When a root cause of an issue is attributed to human error, what does that really mean? It could be incompetence or carelessness that contributed to the issue, or it could be something else. When someone is overworked, they are tired, not able to make optimal decisions and have a much greater chance of causing a major issue. The problem is not only generational, but societal. Nowadays, working to the point of exhaustion is a badge of honor. For people in IT, it is not uncommon to work 60 to 90 hours a week. In some parts of the world, such as Everyday must be jumpstarted with coffee and energy drinks. Working all night to finish a proposal, or perform some afterhours maintenance is normal. Working across time zones means that you may be adding 3 hours to your workday if going from coast to coast in North America, 8 hours from the San Francisco to London, 16 hours from Vancouver to Tokyo.

Working while exhausted has been proven to be just as bad as working while intoxicated. We wouldn't want to have an entire department to be drunk all day, every day, so why is exhaustion something that is aspired towards?

In Europe, the average work week is less than that of North America, whereas in Asia it is much higher. The amount of time allocated to work is one thing, but the true measure of chronic exhaustion is the amount of sleep that is obtained on average over a 24-hour period.

It is possible to work 12 hours a day, 7 days a week and still get 8 hours of sleep every day. There must be a balance of what people are willing to forgo. It is difficult to work 12 hours a day, have an active social life, recreational time, work-out, and spend time with family, while still getting 8 hours a night of sleep.

The minimum amount of sleep required for humans, regardless of region, ethnicity, altitude or hair color is 6 hours of sleep. If less sleep is obtained occasionally, then the body can adapt and compensate. If less than 6 hours becomes a continual trend, then the effect on the body is likened to brain damage and heart attack. The body needs time to recoup and heal. The mind needs time to process and create the subconscious connections and associations. Lack of sleep will impair both short term and long-term memory, decision making and communication skills.

In professional trades, continual lack of sleep can lead to engineering disasters. In the medical profession, it can lead to malpractice and loss of life. In IT, it can lead to lax security practices, lack of standards and policies, and human error, which can lead to data loss, information theft and financial impact to all associated organizations.

Regional culture of overworking can contribute to the possibility of human error, but it is not the sole culprit. There is also the 'toss it over the wall' mentality. This is when there is a communication issue between the different departments. Examples of this are when the sales team makes promises that are impossible to architect, architects design solutions that are impossible to deploy, provisioning deploys a solution that is impossible to support. In the end, operations teams are on-call 24/7, supporting solutions that were flawed from the beginning and they make mistakes because they are burnt out. So, who's fault, is it? The accountability is with the organizational leadership and management. Everybody plays their part, but it is the job for management to see where things are broken in the organization.

7.9.1 How to Protect Against Human Error from Being Overworked

There are many things that can be done that can help stave off burnout, beyond getting an intravenous espresso drip.

Caffeine is the world's most popular and most abused stimulant. In 2011, the world produced 18,651,110,000 lbs. of coffee. Global consumption is estimated to increase by 25% by 2020, as regions like China, India and Latin America become more 'westernized'. However, global supply is shrinking because of a wide spread fungus called 'coffee rust' (Hemileia vastatrix) that is killing off crops. Soon, you will only be able to get expensive, low quality coffee.

7.9.2 Tips for Employers

1. Explicitly define everyone's role and see if there are people that are doing more than what is defined. If they are, then either they do not feel challenged enough, or more staff need to be hired.
2. Provide clear expectations on what is required, measured and evaluated for performance.
3. Ensure that teams work well together and that everyone is pulling their weight.
4. Have frequent open and honest dialog with staff and management.
5. Create goals for individuals and team and review progress at defined schedules
6. Ensure that after-hours incidents are only handled in 4 hour shifts, with a hand-off to the next on-call.
7. Ensure that employees have a path for advancement. Provide training opportunities, incentives and encouragement.

7.9.3 Tips for Individuals

1. Do your most difficult (screamer tasks) first thing in the morning, before anything else. You can get more done in the first few hours than the rest of the day combined.
2. Take regular breaks. Don't work for more than 2hrs straight at a time. Get up. Go for a walk.
3. If you can, spend at least 20% of your work time on something innovative over which you have ownership. It can be something small like creating a new way of inventorying hardware, or automating a process. It will differentiate you as a contributor and benefit the company. You will also feel pride in your achievement. Win-win.
4. Shut down at a certain time. Disconnect from worrying about work and give yourself some time every day to relax and do something for yourself.
5. Spend time with friends and family on a regular basis. If this is not done, you can lose years and relationships in what seems like the blink of an eye.
6. Create personal and career goals for yourself. Track your progress and review it frequently. Socialize what you are working towards with your work, peers and family. Find people that are on a similar trajectory and work together to help each other achieve success.
7. Make sure your work is interesting to you. It is not always possible, but work towards it as a career goal if that is not the case.
8. Work towards improving the skills that you apply at work. As you get better, doors open and opportunities arise where they did not exist previously.
9. Try to limit caffeine and sugar intake. They create dependencies that cause emotional swings that can affect work performance and can ultimately lead to health problems and reduced life expectancy.

7.10 Chapter Summary

1. Understanding the motivations and reasoning behind decisions that people make will help you relate to them and increases the reliability of predicting their future decisions.
2. Understanding the perspectives of others and how a situation is perceived, will help you navigate any future conflicts with greater ease.
3. There are many reasons why people do not act in a logical manner. If you can recognize cognitive biases in others, then you can navigate the effects of them without having to react to them.
4. People are poorly protected, mushy computers. They are the easiest to attack and manipulate. There are no log files, no firewalls and many vulnerabilities. Social engineering capitalizes on this.
5. A disgruntled employee can cause more damage than a bull in a china shop. See the signs and never let it get to that level.
6. Culture is the life blood of an organization. It is also unique and a result of many complex interactions. Take the time to understand it and nurture it. Corporate culture does grow organically, but it can be guided in a positive manner.
7. Synergies are needed for successful acquisitions and mergers. If they are not there, then the relationship will evolve like a cancer. No one will benefit.
8. Incompetence and careless can both be remedied with training, communication and accountability. It is the job of the organization to recognize it and address it. If it is not addressed, then it's just a matter of time before it causes a major incident.
9. Working too hard and not having an effective work-life balance will have a great financial impact on an organization and health impacts on employees. Overwork should not be a badge of honor.

7.11 Chapter Review Questions

1. You have the option of choosing 1 of 3 WAN service providers for your organization. The requirement is to have the highest availability for the lowest price.

	WAN SP A	WAN SP B	WAN SP C
Cost	$200 / month	$500 / month	$175 / month
BW	100Mbps	1Gbps	1Mbps
Avail	99.9%	99.99%	99.99%

SP = service provider
BW = Bandwidth
Avail = Availability

Using a decision tree, map out the best option based only on the requirements.

   a) If the requirement was to have the best balance of price / bandwidth, would this change the best option?

2. A sales executive talks to a customer about a technology that they read about online last year. There has been a major release of the technology since then, that the sales executive does not know about.

   a) What is the cognitive bias of the sales executive in this scenario?
   b) What is the risk due to the cognitive bias?

3. A knowledgeable Systems Engineer accidentally deleted a production database during a maintenance operation.

   a) What are two possibilities that caused or contributed to this occurring?
   b) How can this be prevented from occurring in the future?

CHAPTER 8

Modeling Outcomes

The best way to predict the future, is to create it.
--Abraham Lincoln (Former President of the United States)

To truly understand the effect of a single decision, you need to map out the dominoes and see what they touch when they fall. This chapter provides you with the tools to model risk scenarios and gage them with Key Risk Indicators (KRIs).

8.1 Modeling Risk Scenarios

In the 1980's science fiction writers like William Gibson and Bruce Sterling, helped create a genre called Cyberpunk. The stories took place in a dystopian future that could very possibly occur with the trajectory that the world was taking. They called out risks, and concerns with technology and society and were often fraught with governments led by artificial intelligences, cybernetic implants and flying cars.

30 years later, we still don't live in a world like that of Blade Runner. Or do we?

We have videoconferencing capability on all smartphones, people are talking at conferences with robotic telepresence, AI personal assistants are commonplace and we have self-driving cars running on sunlight.

The list of "future-stuff" that is here and now could go on and on. We take our everyday lives for granted and it's hard to even imagine a time before the technologies that we see as commonplace even existed. Well, futurists could foresee it and science fiction writers could imagine it, so what did they do to envision these possible futures? What was their process?

They created scenario models based on key factors, then expanded on them in multiple directions.

A risk scenario model is like a story that occurs 30 seconds in the future. It is very possible that it can occur, but it also may not. A specific sequence of events needs to be defined for a scenario to be comprehensive.

> *"In a sense, if you're not getting it wrong a lot when you're creating imaginary futures, then you're just not doing it enough. You're not creating enough imaginary futures." –William Gibson*

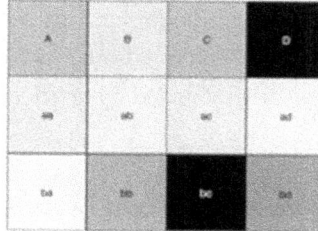

Figure 8. 2 - A visual representation of an environment state. The shades of grey correspond to numerical values of 0,1,2,3

These sequences of events are state changes. First, inventory all the moving parts: technologies, people, processes, adherence to policies, etc. Then map out all the states of these parts and you get something like the diagram below:

Inventory items are mapped out in shades of grey. There is a total of 4 states here (0,1,2,3) for each inventory item. Each row represents a different area, such as technology, people and standard processes.

Using some calculations, it is determined that the total number of states for the environment is the number of possible states to the exponent of the number of inventory items.

So, if we have 12 items in inventory and 4 states for each, then the number of possible environment states is 4 to the exponent of 12 or 16777216. This can be identified by:

1. A visual diagram.
2. Calling out the individual states of the items.
3. With an ID.

Compare the different modes of representation.

In a visual diagram, we would see this:

If we represented the individual states, it could be represented like this:

A-3,B-2,C-3,D-4,aa-2,ab-1,ac-2,ad-1,ba-1,bb-3,bc-4,bd-3

An ID could be represented in many number of ways, depending on your preference.

The ID could be 12-4-323421211343 which would identify the number of items, followed by the number of possible states, then the actual states for the inventory items in descending order.

Or:

1-ab, ba, 2-B, a, ac, 3-A, C, ad, 4-D, bc which would identify the states, followed by their item identifiers.

Or:

The items can be coded in bits, 2 bits for 4 states.

0 = 00

1 = 01

2 = 10

3 = 11

The number of items will represent the binary string. If you have 12 inventory items, then you have 24 bits

323421211343 would turn into 100110110100010000101110

You could then turn the binary into hexadecimal. The hex environment state ID would be 9B442E.

Using straight binary, you would need a map to what the item is. So, you could have something like this:

Technology = bits 17-24

People = bits 9-16

Process = bits 1-8

Let's define some inventory example items for technology:

A = Storage Latency – states (0=All paths down, 1=Extremely high latency, 2=high latency, 3=optimal)

B = CPU Resource Contention (0=CPU halt, 1=high wait, 2=high run, 3=low usage)

C = Memory Resource Contention (0=paging, 1=ballooning, 2=high utilization, 3=low utilization)

D = Network Load (0=packet loss, 1=high latency, 2=high load, 3=low load)

Now let's define some examples for people:

Aa = Operations capabilities (0=team is as bright as a bag of hammers, 1=limited skillsets, 2= team functions and can address most demands, 3=SEAL team sixe* can take some pointers)

Ab = Resource utilization (0=team has stopped sleeping months ago, 1=everyone is on-call and all vacations are cancelled, 2=work is steady but still time for personal improvement, 3=team has been working remotely from Vegas for the past few months)

Ac = Operational responsiveness (0=all requests for support must be faxed in, but number is out of service 1=operations team is just a flash mob made up from all other departments, 2=real-time alerts and service dashboard available publicly, 3=proactive operations that align with strategic resiliency plan

Ad = Ability to adapt and learn (0=extremely rigid and will not attempt to keep up to date, 1=upgrading skills requires organization mandate, 2=team keeps relatively current with technology, 3=mirrored production development environment used for testing new technologies and staging / testing before moving to production)

> SEAL team six is an elite counter-terrorist, hostage rescue and reconnaissance task force

Finally, let's define some process items:

Ba = Change management process (0=no change management, everybody makes changes, 1=no change management, specific teams or responsible for specific technologies, 2=change management database exists, but only documented after the fact, 3=full change management process with change advisory board, CMDB and scheduled CAB meetings)

Bb = Project management leadership (0=no project management, no technical strategy, 1=technical roadmap exists but individuals responsible for

Bc = SLA adherence metrics

Bd = Operational onboarding process for new technical staff

These are only 12 inventory items. If we were a bit more detailed, we may have 100 inventory items. If they all had 4 states each, then we would have a 200-bit number. This would convert into a 50-character hex number. To simplify the calculation, you can think that every 2 inventory items will equal 1 hex character (with a 4-state system per item).

8.2 How to define KRIs (key risk indicators)

Key risk indicators, or KRIs are used to determine if there is a change in the probability or impact of a risk. If a KRI changes, then it will influence the criticality. The more KRIs that are mapped out, the more reliable the measure of probability is. More inputs provide additional information with which to make decisions. When the change in criticality occurs in a risk, this will change the overall priority of action that is taken to address it.

For example: a risk that has been accepted because of the low impact and the high amount of effort to mitigate it, may become public enemy number one and on the center stage given the activity of KRIs.

A scenario for this would be:

Infrastructure component – Environmental:

A datacenter requires cooling to ensure that equipment won't overheat and fail.

Risk:

Not having the air intake for the HVAC protected from animals, because they could cause damage or block airflow if they entered the conduit.

KRI:

Squirrels are becoming more active on the roof. This is an indicator that they may have made a nest in or around the roof.

KRI:

HVAC system is not able to provide the same level of cooling for this time of year as it did in the past.

KRI:

Filtration system needs to be cleaned more often.

KRI:

Bird droppings are completely covering everything in the air intake area.

KRI:

Datacenter staff are not doing outdoor generator checks as frequently because of the mess left by animals.

KRI:

HVAC failures are occurring.

Initial probability for the risk was scoped at 3/10 because there were no signs of any animals in the area and it seemed unlikely.

Initial impact was scoped at 8/10.

Criticality (P x I) was therefore 24.

Figure 8. 4 - Talk to the people closest to the infrastructure

After observing the KRIs, it was determined that there was a level of animal activity near the air intake that has started to impact the HVAC system.

Probability is therefore changed from 3 to 10 (because it is happening, not just a possibility), which changes the criticality to 80.

Some KRI's can be obtained from monitoring metrics, others from information that is aggregated and interpreted like a news feed. Sometimes the information can only be obtained by conversations with other people. The latter method is the most difficult to scale, but may provide a wealth of information that could not be obtained otherwise.

A good example of getting KRIs from people, would be to talk to users, frontline and operations staff. The people who are closest to the technology are not always the ones that architect it. A good feedback loop is to keep an open dialog with them so you know when a mole hill will turn into a mountain. Another example is to talk to industry subject matter experts and people of note in the community. This is what investors do to determine where to put their money. This is also how you can determine where to put your resources and time.

8.3 Modeling Future States

You now know how to map out your existing infrastructure state to a hexadecimal code. You have mapped out the risks in every part of your environment with visual models and created a risk register. Then, through monitoring KRIs, you get a good estimation of the probability for each one of those risks. What now?

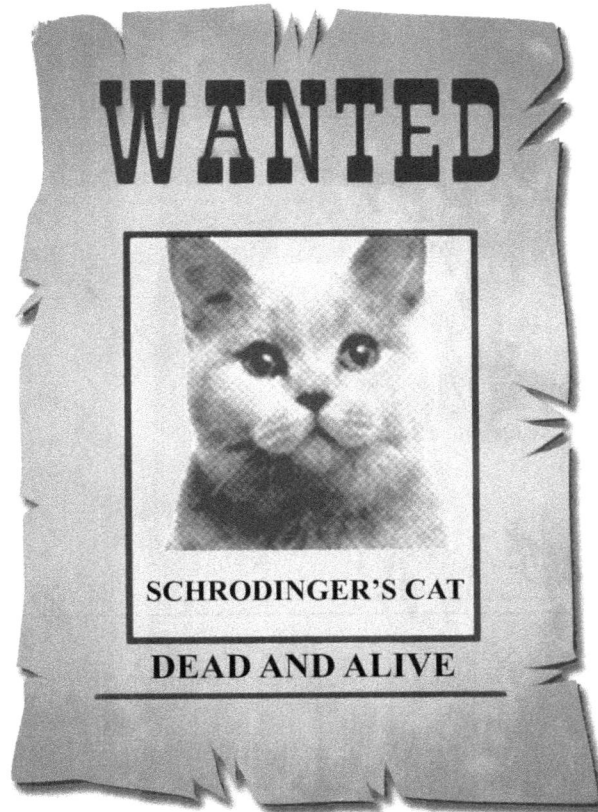

Figure 8. 5 - Schrödinger's cat represents all possible states existing simultaneously until one is observed

This is where you do your fortune telling and your stories unfold. What may have started out as "Lorem Ipsum dolor dolor sit amet" (placeholder text), now must become "It was the best of times, it was the worst of times." (Charles Dickens; A tale of Two Cities). Ideas must be laid out and unfurled to their ends.

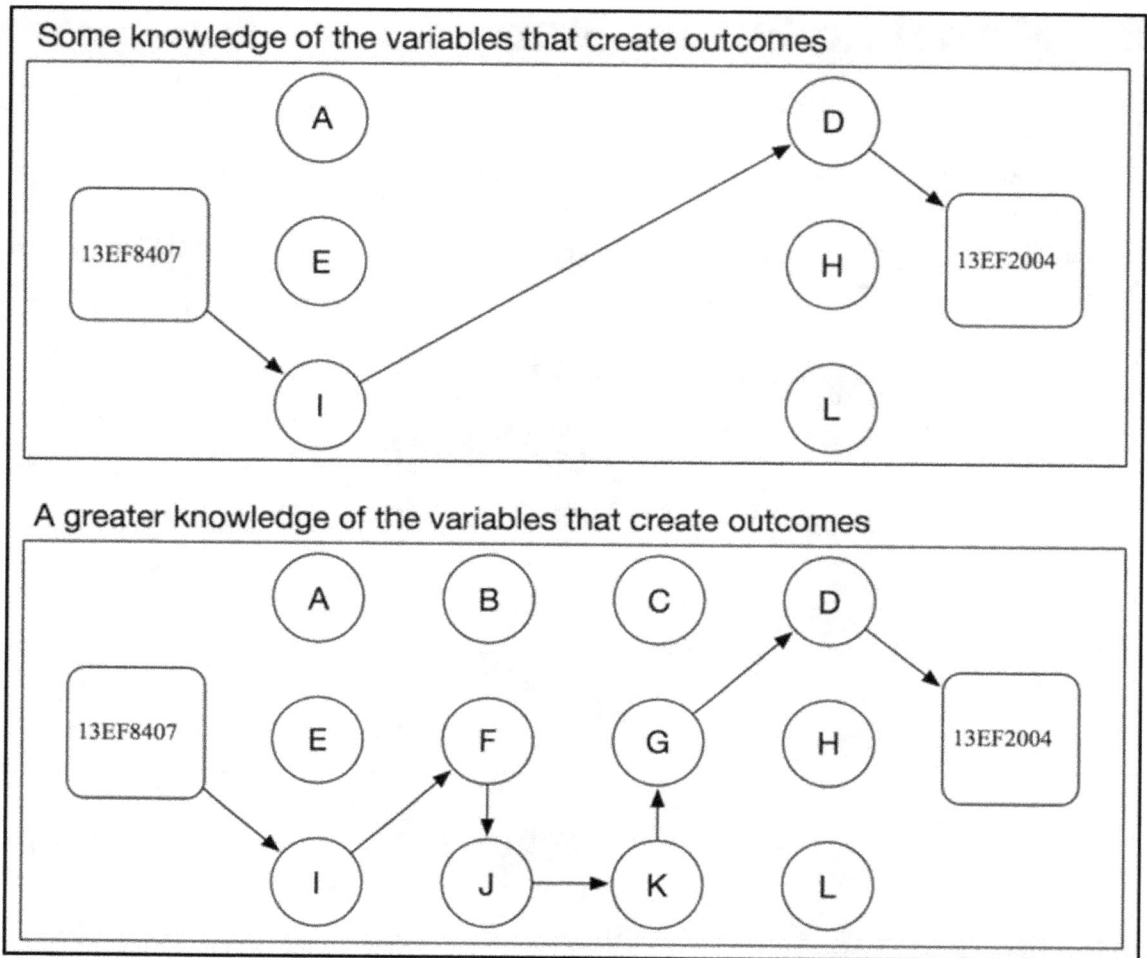

Figure 8. 6 - The frequency of state change sampling will determine the detail of patterns observed

In quantum physics, there is the "many worlds interpretation" or MWI, that states that every possible history and future occur in an infinite number of parallels worlds. We can only perceive one at a time, but they all exist simultaneously in the realm of possibility. Just as Schrodinger's cat is both alive and dead, all possibilities are real until they reveal themselves. The possibilities that occur are the ones that are often the most likely.

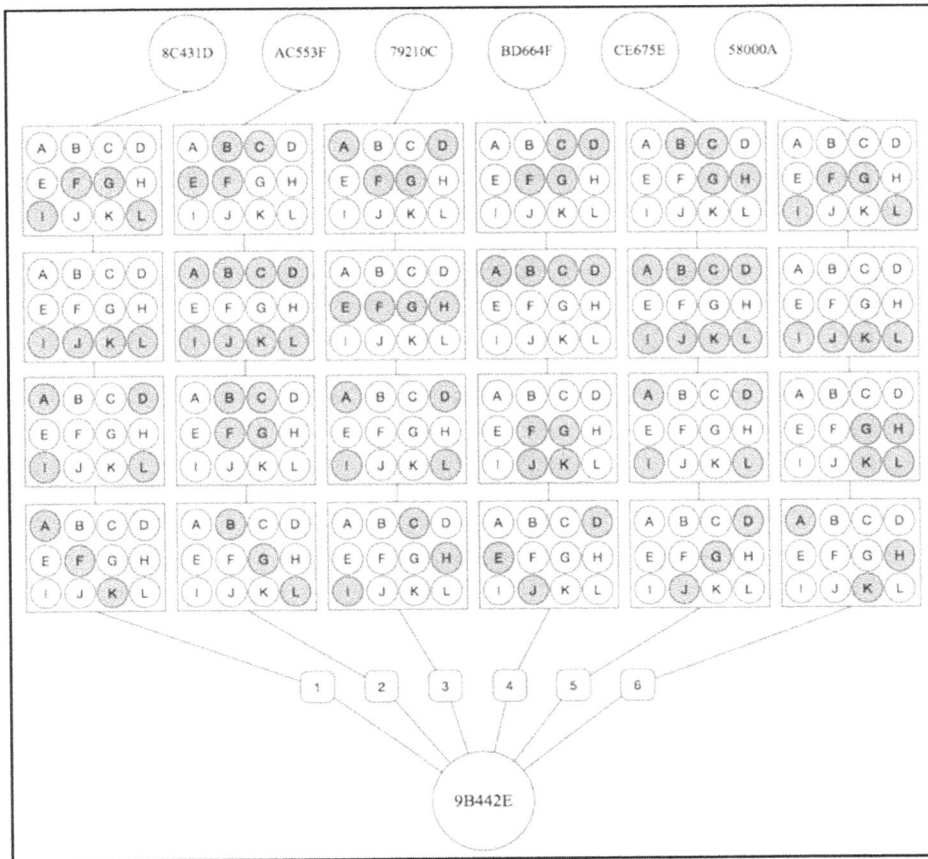

Figure 8.7 - Mapping out possible state change patterns and their outcomes

However, the only way you can know what is likely, is to know as many of the variables as possible that will affect the outcome. If a current state is, for example (in hex) 13EF8407 and it will become 13EF2004, then what are the causes for each change that occur in between?

What are the variables, the catalysts, the interactions? In the diagram to the left, these are what is represented by objects A-L. 2a

In the top example, only a limited number of variables are known. The events that occur may be large, or they may be micro-events with ripple effects. The more events that you can pinpoint between these two states, the better your prediction will be. No prediction is perfect, so it must be tempered with alternative possibilities. A future model map may look like this:

This diagram is a little more complex. With the previous diagram, we talk about variables and how knowing what they are will help us determine the state changes that occur. In this diagram, we see a path from a beginning state, to a final state. In between are a series of state change events. Each one of these events is the "variable" from the previous diagram. This diagram shows the actual state effects of the event / variable.

The effects in the states that are caused by an event are not represented in a linear manner. They are represented in a sampling frequency. A state change due to an event may occur in a single sample change, or it may occur across many of them. This makes it extremely difficult to correlate exact state changes to specific timelines in a state change from an event. Sometimes it is not the details that are important, but rather the picture they paint. If you see patterns reoccurring, then there is something there to look at. In each of the state change sets that are shown in the diagram above, there are very similar yet different effects of events occurring.

In the mathematics of complex numbers, there are methods of visualizing infinity such as the Mandelbrot or Julia sets, which make fractal patterns like the image to the right. In each of those examples, a property they have is self-similarity.

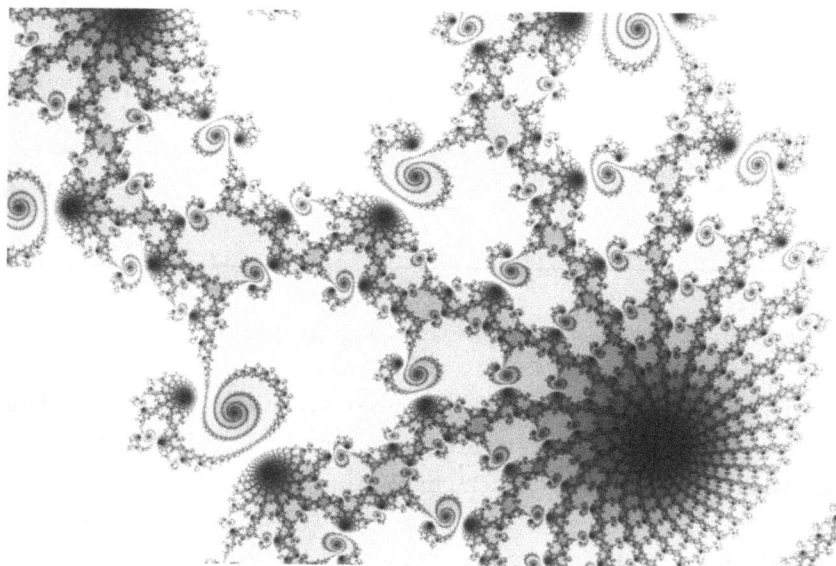

Figure 8.8 - The visualization of a fractal set

This means at certain levels of iteration, they will repeat in a very similar way to how they have in the past. It is this property of self-similarity that we will be looking for in the state changes.

If we can see the patterns and the time it takes for the iterations of the state changes in the environment, then we can make calculated guesses about what and when state changes will occur. Let's look at a pattern that is a bit more concrete and relevant to what we are talking about.

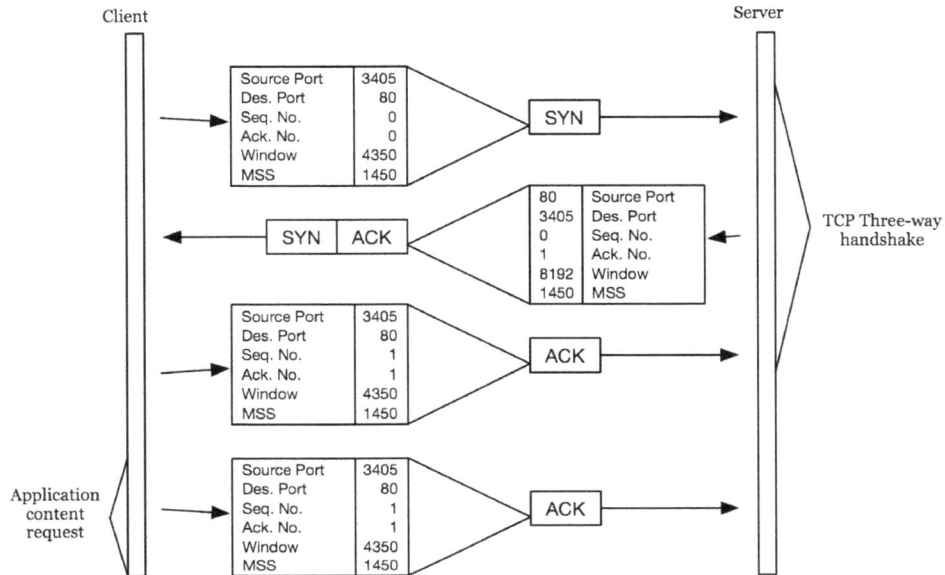

Figure 8.9 - A TCP 3-way handshake

This is a TCP 3-way handshake; SYN, SYN \ ACK, ACK. It has a very definable pattern when you look at it in a network packet capture. This is the same for many other TCP and UDP sessions. By seeing the patterns, you know what comes next. If your savvy enough, you can determine the causes for each pattern that you find in a packet capture and associate it to a source event. It may be very difficult if there are several events contributing to the noise and deciphering the sources from that is almost impossible. In those cases, you need to simply see the pattern and recognize the potential effects, so you can prepare for them. You may not have all the pieces of the puzzle, but you will likely have enough to make a considerate guess.

This is the reality that needs to be pondered; how can we guess what every micro event is and what its impact will be on the infrastructure at hand. The truth is, we can't; though we can make some observations and infer their effects and directions. Below are some suggested steps for future modeling:

1. Pattern recognition.
2. Measuring trajectory.
3. Understand cycles and trending.
4. Interpret heuristics.
5. See the next state (Tetris block).
6. Collect radicals.
7. Devil's advocate.

8.3.1 Pattern Recognition

Patterns occur everywhere, they are the building blocks of physical reality. A sunflower has a very specific pattern in its seeds that is known as the golden ratio, or the Fibonacci sequence. It has many of these interwoven together because it is a naturally efficient design. When a system is designed and it is given parameters by which to operate, the details of how it performs are not always acknowledged by its designers as much as the end result. This indifference to anything but the result is self-serving as it allows for focus on the core functionality of a system. This is important, but once that functionality is there, the unintended design needs to be looked at.

If you fill a balloon with helium and release it, it will float into the sky. This is a good example of building a system and having parameters for operation. The intended result of this system is to carry a rubberized spheroid into the sky. An unintended design is the fact that due to wind currents, the balloon will follow a path that may seem random. However, it is being guided by the global weather system, which is affected by ocean tides, interplanetary positions and the heliosphere weather conditions of the sun.

No one is omniscient with the ability to see all the interrelations of all components (intended and not intended) of an open system (in this instance, I am referring to an open system as one that interacts with the physical world in an unconstrained manner). However, the more closed a system is, the more likely that given enough time, one would be able to understand and map out many interrelations and the patterns that define their interaction.

8.3.2 Measuring Trajectory

If you have identified a pattern, the next step is to define the cycle and the steps within it. If the pattern A, B, C, D occurs and every time D occurs in the cycle, it causes "x to do y", then you can identify that pattern by its effects. This allows you to plan for "x to do y" by recognizing the initial start of the pattern. Which in this case, is A, B, C.

An example is this sequence of events:

A manufacturing company makes fireplaces from design and testing to construction and delivery. The design team has CAD software for rendering models of fireplaces and they also perform heat simulations with their analysis software. Both of those tasks generate very large files that need to be archived on a file server. The designers would work until 4pm, then upload the data to the file server.

Some days the files would not upload correctly and the network would become unreliable. This became an issue for productivity, but instead of resolving it, they worked around it. They would upload their data in the morning the next day, or simply keep it on their workstation. Productivity was back up to speed, but the root cause was never resolved.

A while later, backups of the local data repository on the workstations had to be done because there was the risk of data loss. The backups ran in the evenings and archived data on the workstation would work most days, but not on Thursdays or Tuesdays. On those evenings, the network would become unreliable and all communication would get lost.

The issue was found to occur from 4:30pm to 10:30pm. All connectivity would be lost for the designers. The users compensated for this and learned to work with it like a three-legged dog playing Frisbee. It wasn't as elegant as it could have been, but they could still get the job done.

> This was a real case. The case of the missing packets! – It was determined that the unshielded CAT5 cabling that was going to the designers' offices passed very near to an arc welder on the manufacturing floor. The electromagnetic energy discharged by the welder caused enough EMF interference to completely take out the network equipment, which was in a continual state of rebooting if the welder was operating. The timing of the occurrence was due to a certain shift worker using the specific welding equipment in that area.

The trajectory was this: network services for the designers would go offline on Tuesday and Thursday 4:30pm-10:30pm. You could expect backups to fail if connectivity was lost.

If connectivity was lost, then it would likely be lost for up to 6 hours. The users identified the pattern, accepted it and worked around it without knowing the root cause.

8.3.3 Understand Cycles and Trending

In the above scenario, the problem that the users experience happened at very specific times because of a scheduled recurring process (the arc welder shift on Tuesdays and Thursdays). If something happens at the same time every week, it can be easy to plan for. However, if the cycle is not as repetitive, then it can be more difficult to pinpoint.

This is when you must look for trends, events, process workflows, understand the weekly business cycle, the monthly cycle, the staff work schedule, the maintenance routines, patch Tuesday, exploit Wednesday, etc.

Create an infrastructure calendar and map out every event that occurs over the month and year. Give each event an ID and map out the effect it will have on the environment, the risks it will have or create and any strategies or actions that could be performed to lessen the overall impact.

Minutia and detail are not important for the calendar items itself, because they are just used as a tool to quickly visualize what is happening and what the effects are. The details can be housed in a change management system. Every technology has maintenance schedules, cycles in operations and lifecycles. Create the calendar by going through each technology in your infrastructure, map all events, then do the same for applications, then layer on everything from the change management system.

By knowing the standard operation cycles, how those look, and the expected effects of each event, you will know if a deviation from the norm is occurring at any given time.

8.3.4 Interpret Heuristics

Heuristics are essentially a means of machine learning and artificial intelligence. It is often used in systems that automate the process of returning it to a desired state when a deviate occurs. These heuristic systems will ingest raw data from logs, event systems, etc. then use a discovery or monitoring tool to understand what the sources of the data are. A predictive engine will know what to expect based on the data sources, the events that occur and their schedules. However, desired states have been programmed into this system as the state to be. Thus, the system will use its analytical engine to determine the best ways in which it can achieve a return to the desired state. The methods it determines can be automatic or suggestions for a manual operator to perform.

What these systems will not do, is provide the reason why something that was not expected occurred. They can provide information on when the unexpected event occurred, its effects and the recommended action to fix it, but not the root cause.

By performing a root cause analysis on the events in an environment that cause a deviation from the desired state, you can better understand the actions that may need to be taken to alleviate the frequency of deviation, or remove it completely.

Going through this process will allow for prediction of deviations based on historical precedent and probability defined by the known variables in the environment.

8.3.5 See the Next State (Tetris block)

In the game Tetris, while you are manipulating falling blocks, the next block that you will have to deal with is presented to you on the top of the screen. You can do one of two things with this information:

1. Ignore it and deal with the problem of the current falling block.
2. Use that information to influence how you manipulate the current falling block.

When certain events occur and you know the transition of states that will happen based on historical precedent, then you have a window into the future (however short it may be). This can be done by mapping out the state transitions that occur from a scheduled or unscheduled event. If either the exact same state changes occur the next time the event happens, or if it can be replicated, then you have a valid mapping.

Now each time the state change pattern occurs, you will have the "next block" available to use or not use, depending how you play the game of IT operations.

8.3.6 Collect Radicals

Have you ever seen a guy on the bus with a mischievous smile on his face, eating a sandwich, with no shoes on his feet, hands covered in paint and pockets full of cash? You know there's a story in there somewhere and it's going to be a good one, but you'll probably never find out. That guy is what is referred to as a wild-card, or a radical. You come across them occasionally in life and probably more often with the equivalent in technology.

Do you remember that time when a network port started flapping uncontrollably, or when you could ping an IP, but all services on that machine were inaccessible and there was no firewall? Or that time when a Java app decided that it needed all available memory on a server to run a simple GUI interface?

You can call these things many names: runaway processes, bugs, design flaws, operation aberrations, gremlins or even bad juju. These are the wild-cards, the radicals, the things that drive users to smash things and drive admins to drink. However, they are also treasures in disguise. Find the stories, collect the anecdotes and use these in plotting events when modeling outcomes. You may have planned for power failure, earthquake and sabotage, but did you plan for the 1000-gallon fish tank two floors up smashing and somehow leaking into the ventilation system?

Figure 8.10 - A user is frustrated by their "buggy" system

Did you plan for the extremely contagious flu that happened to affect the entire operations team a day before a major migration of services between datacenters (with tight timelines)? What about the faulty power supply in a server that smoked out the datacenter and caused the sprinklers to go off because the FM-200 fire suppression system was under maintenance?

Think of the radicals as IT war stories with morals.

8.3.7 Devil's Advocate

The best laid plans of mice and men often go awry. This is the best phrase to explain this method. It is used in change management to say, "What if your plan goes off the rails?" What is your backup plan? Then in the change request, a detailed failback strategy would be in place to revert to a rollback state, or a path to overcome the issue with a pre-defined strategy.

In this instance, we are taking it a little further. We would be diving into the detail of how something goes off the rails. One method of doing this is through table-top exercises. This is a method of having a real-time train of thought on possibilities with several teams that may work in silos. It goes something like this:

1. Get the entire operations team in a room for 2 hours.
2. Get food. People are more open and talkative when there is food.
3. Provide the scenario background. Things leading up to the current state, drivers, direction, infrastructure risks, etc.
4. Provide an event that changes the state and what the desired outcome would be. An example of this would be:
5. An organization with multiple sites has a single site that has its own separate union. The union at this one location has gone on strike. Some production services run there and there is a risk of service outage. What does each person in operations do during this process to ensure services are not impacted?
6. Do a round table and get everybody's response.
7. Have the teams build on each other's responses until you have multiple strategies for a desired outcome.

Congratulations, you just crowd sourced your modeling of future states! Make sure you have a very detailed record of all conversations that occur during this process. You will want to review them and expand on some areas and the concerns that were brought up, but not yet fully explored.

8.4 Reducing Possibility

A philosophical razor is something that allows someone to remove unlikely explanations for an event or phenomena. Some well know razors that can be used in the modeling of future states, have been paraphrased below:

1. Occam's razor: The simplest explanation is often the right one.
2. Halon's razor: Don't assume malice if something can be explained by stupidity.
3. Alder's razor: If something cannot be determined by experiment or observation, then it is not worth discussing.

The Australian mathematician, Mike Alder created Alders razor. Its alternative name is called "Newton's Flaming Laser Sword".

The concept is that the observable universe becomes much vaster when you incorporate personal feelings, the seemingly impossible, or the unprovable.

This removes things such as opinion, politics, religion and philosophy from conversations that require fewer variables. Newton never had a Flaming Laser Sword because of many factors, such as the limitations of science at the time and the fact that it would serve no purpose. However, it would be awesome as a plot device in an "age of enlightenment" science fiction story.

Figure 8.11 - An artist depiction of Newton's Flaming Laser Sword.

8.5 Intra-Organizational Politics

Understanding an infrastructure and its limitations is not solely a technical endeavor. The capabilities of an organization are defined by how it operates internally. What are the internal politics between departments? What is the process for sourcing and acquisition of equipment? How is the budget defined? What tensions exist between departments, personnel or vendors? Understanding these will subtly sculpt how approaches to solutions are thought out.

Start with organizational charts, and then learn the roles and responsibilities for everyone that directly or indirectly affects IT. Learn the process workflows, the vendors of record, RFP process. Talk to people within the organization that have gone through a major project. Ask them about the battle scars, the one that got away and any other stories that they are willing to share that give insight to internal politics.

Some organizations are subsidiaries of other companies and although they may be independently operated, they often have a mandate to standardize across the greater organization. This allows for purchasing agreements and support that are done on economies of scale. However, this also means that many decisions and processes are in the control of the parent company.

8.6 Market Pressures

When the market turns in a certain direction, it affects the way things get done and the pace with which it happens. An economic downturn or a financial loss could force an organization to cut costs very quickly. An opportunity that arises could force a change to be implemented on an accelerated timeline. When these things happen, it limits the possible paths to the few that must be followed.

So, by knowing these external constraints, you know the probable outcomes and have an idea of how they will occur.

8.7 Financial Viability of Risk and Return on Investment

When a company builds a service offering, whether they advertise it or not, it is designed to a certain level of availability, redundancy and resiliency. It doesn't matter if it is on-prem, in the cloud or hybrid. The amount of money allocated to these three things will determine the level to which they are built up. When there is an outage, how much does that cost versus the capital to ensure that there is not an outage? What is the public perception of an outage? When many organizations use a cloud services provider as their core infrastructure, then many companies will be affected when there is an outage of that service in a region. If the company that is running their infrastructure in the cloud has redundancy across multiple regions and they have a load balancing mechanism, then a regional failure will not impact them in a critical manner. Companies that do not have multi-regional load balancing will be critically affected. However, if enough companies are experiencing outages due to a cloud services provider region failure, then it is easy to blame the provider and not the companies for a non-redundant design.

The risk of an impacted public perception of the company is reduced by transferring the blame to the service provide and hiding in the chaos that ensues from the outage. What is the user tolerance threshold to an outage, or multiple ones? What is the cost of redundancy to the bottom line? What is the financial impact of an outage? What is the mitigation strategy to reduce the financial impact versus removing the risk?

A design can be guided almost entirely on economic constraints instead of technological ones. The end design is not necessarily pretty, or optimal from a performance or availability standpoint, however it will make financial sense. I do not recommend that the finance department have too much control over the design of the infrastructure without having a solid understanding of the implications of every decision. It is the job of the architecture team to provide this information and present it in an easily consumable manner. Ideally, the budget constraints are understood by the architecture team and used to direct the infrastructure design.

If you want to be ruthless and do everything in the cheapest way possible, then you need to have a greater understanding of the risks and ramifications of your decisions. A good example is a Bitcoin mining operation. These organizations try to get the maximum return on investment from the cheapest methods possible. They are often with compute clusters that are run in massive warehouses, on shelves, without cases and with open-air cooling and cables running everywhere. Power is often the biggest constraint in these operations, thus an electrical engineer should be involved in the design and deployment of the power to the datacenter. If they are not, then the operations team risks power failure, or worse, such as a total facility meltdown from a fire. Where are things skimped on? Fire suppression? PDUs? If these have the chance of failing because you're being cheap, then you need to spend more time on thinking how to compensate for this. How creative you are when you're planning on being cheap, will determine the cost for the ruthless technology cuts.

8.8 Environmental and Regional Considerations

Every region has its own climate and environmental cycles. Whether its tornados in the mid-west, earthquakes near the continental fault lines, or forest fires or floods, every region is different. Once you know the regional environmental patterns, then you cannot only approximate the likelihood of an occurrence, but you can also see what the historical impact has been.

If an office or a datacenter that you manage is prone to events such as the following, then you must plan for the damage and recovery:

- earthquakes
- hurricanes
- tsunamis
- flooding
- climate change
- political stability
- poor air quality index
- forest fires

The local government of the region will likely have a community impact map for the likely natural disasters. Use these, as well as historical references, to determine the impact that may occur and the actions that have been taken in the past. This will help build future models that will influence how a business continuity and disaster recovery plan respond to real world events.

8.9 How to Keep the Lights on When the Company is Doomed

50 percent of small businesses fail within the first five years and 40-50 percent of the U.S. work force is made up of small businesses. This means that many people will inevitably work within, manage or own a company that will go under. There are many reasons why this would happen: market saturation, product or service demand change, regional economic climate changes, management or staff related issues, rapid fluctuation of revenue sources, mergers, acquisitions, etc.

I have personally worked for 6 different companies that have decided to pack it in. None of these were due to mismanagement, but rather to external pressures and the economic climate. People respond to stress and uncertainty in different ways. This is often due to personality, financial stability and obligations to other people. A person that has worked in the same role for years and has resisted change, will find the prospect of leaving it unnerving. Some people may be proactive and train up for better chances in the job market, whereas some may wait it out and see where the pieces fall. Sometimes, people hide in the cracks of an organization until they can do that no longer.

I remember one organization that was thriving when I started, then started on a rapid and irreversible decline. This company was telecommunications oriented, with a huge focus on business and residential phone services. They were a successful medium sized organization making hundreds of millions of dollars a year. However, because of the decline of people wanting or needing a land line, the rise of smartphones and the millennial generation refusing to talk on a phone, the climate had changed and the business model was in decline.

User attrition was occurring at a rate of 25% per year and no amount of good marketing or bundled deals could save it. Ironically, my wife and I got rid of our landline just before I started working at this company. The signs were everywhere that the company was going under and to their credit, they acknowledged this. The goal for management at this point was not to save the company, but to slow the bleeding. This consisted of:

1. Budget cuts.
2. Wage / bonus freezes.
3. Benefits cut or reduced.
4. Staff cuts.
5. Closing offices.
6. Selling off assets.

7. Splitting off parts of the company, polish, shine, sell and repeat.
8. Reduce service offerings internally and externally.
9. Moving to SaaS based services in a feature-tiered model.
10. Acquisitions (counterintuitive, but it made sense).
11. Consolidating departments and resources.
12. Freeze on technology acquisitions and support renewals.
13. Implementing open source technologies where possible.
14. Reviving decommissioned hardware for non-essential services or roles.

Figure 8.12 - When things get bad and some self-surgery is required

These are only some of the methods that were used as a financial tourniquet to stop the bleeding of money.

It reminds me of that scene in just about every action movie, where the hero / heroine gets injured, does some sort of self-surgery with a lighter, a stapler and a bottle of vodka. They then pick themselves up and get on with their day. That is essentially what a company must do once they know that they are going under and there is no return.

Let's talk about what people normally go through when their day to day is forcibly changed and uncertainty is in the air. We will use the Kubler-Ross model, which is known by the acronym DABDA. These are the five stages of grief that someone may encounter when the inevitability of a company's downfall is eminent.

8.9.1 The Five Stages of Grief When a Company is Doomed.

1. Denial

There must be some mistake. I'm sure it's not a big deal and it will get sorted out soon. It's not going to affect me. These are some of the things that go through the minds of people in the first stage of acceptance. It's like when you're at the beach and someone yells, "Shark!" and nobody gets out of the water. Except in this instance, the shark is loss of employment and the one yelling is logic and reasoning.

2. Anger

Once someone realizes that it is not a mistake and yes, this is really happening, then anger sets in. "It is mismanagement and incompetence!" "They don't know what they're doing!" "Why me?" These are the things that become top of mind in this stage. Some people act out and try to "get back" at the company, while others just play the blame game.

3. Bargaining

In this stage, there is a small glimmer of hope, for a short period anyway. A sort of reprieve from the reality that this is really happening. The company is taking a swan dive. "What if we just do this?" Or: "Maybe we can still turn this around?" This stage is kind of like an eye of the hurricane.

4. Depression

In this stage, the math is being done and hope is all but washed away. "Why bother? It's all falling apart!" Some people may become recluse, less engaged, and less social. The motions of work are being done, but they are done lifelessly and without any drive, passion or effort.

5. Acceptance

I heard a story about a boy that fell off a boat in the Niagara River. As he was being swept downstream, he recounted that he went through all of these stages described above. Just before he went over the edge of the precipice of Niagara Falls, he came to terms with his own mortality. He accepted that whatever happens, will happen, and he is just along for the ride. It was in this final moment as he flew over the edge, that he had no fear and it was like a weight off his shoulders. He was clear of mind as he descended through the mist to the unknown waters below. He didn't remember the fall as much as the swimming to a nearby boat afterwards. He lived through it and was changed in the process. He now understood the impermanence of everything in a visceral way.

When a company is going under, the best thing that can be done is to help those that have been along for the ride to something new. Recommendations, referrals and job leads are all things that will help people through the process. A company should be straightforward and honest, hiding nothing from employees at this point. When the stage of acceptance is reached, then employees can make a more effective transition or they can stay until the doors are permanently closed, and the lights are shut off.

On the Titanic, the band played music as the ship sank. The efforts did not stop it from sinking, but it did make the trip to the bottom of the sea more pleasant. If you want to keep the lights on of a doomed company, then you need to get as many people off the sinking ship safely and just have the people that make the music left. The people that have accepted the swiftness of the river and the impermanence of the fall. This process can weed out some great people with a willingness to do what it takes and perform at a high level with no ego or misconceptions of where things are.

8.9.2 Techniques to Delay the Technological Failure of a Doomed Company

When a company has an indication that it may go under, these are some things that can be done to help the people and keep the lights on as long as possible:

1. Employ longer term defense in depth and graceful degradation.

When something fails, plan for it not to get fixed due to financial constraints. The infrastructure design should be able to withstand failure after failure and still be operational, albeit at a reduced capacity.

2. Put experienced management in charge of decline.

If you have people that are experienced enough to manage the IT team under these sorts of pressures, it makes all the difference in the world. You would be more comfortable going into battle with a weathered general, than a freshly minted officer out of training school. The same is true in IT, except the war is with time and money. Like a chess master, IT management needs to be able to make logical decisions to draw the game. Drag it out as long as possible, because in this instance, the game cannot be won.

3. Heavily cross-train on all silo systems.

When the staff is reduced to a skeleton crew, everyone must be able to do every position. Nobody should have information that only they know. Destroy the silos of information and responsibility. Ensure that a mentorship plan is in place to transfer knowledge from one admin to another.

4. Make all IT staff become jacks of all trades and remove specialization roles.

All IT staff should know every aspect of the environment in great detail. The best way to ensure this is to rotate roles and responsibilities. Define a set schedule to rotate weekly.

5.   Increase compute by PXE booting extra computers into a desktop compute cluster.

This one I came across with a company that had part-time contract based workers and massive fluctuations in capital. They were in the animation business and had time sensitive contracts that demanded a quick turnaround. They could not justify the capital costs of getting dedicated equipment and a cloud based solution that did not have the horsepower to get reasonable results due to bandwidth limitations. So, when staff contracts ended or when people went home, they would reboot the powerful desktop workstations and PXE boot them into the rendering farm as nodes. It wasn't a comprehensive solution to their problem, but it did help them meet deadlines that would not otherwise have been possible.

6.   Decommission the most power-hungry systems from operation.

I would tweak that saying in this instance to, "With great power comes great utility bills."

When trying to cut costs, look at your power consumption. RAM is the biggest user of power in a server. Can you reduce the RAM without impacting service availability? If you have 2 CPUs, can the server perform it's roll with one? If the servers are booting of a pair of hard drives in raid 1 or raid 5, can that be changed to boot from SD cards or boot from the network or from a SAN?

Can some of the workload be moved to newer or fewer servers running at a higher utilization? The extra servers can be powered down through out of band management, until they are needed in case of a failure. The infrastructure workload will not be increasing, so you won't need to scale up any time soon. Standby redundancy improves the ability to recover from a failure and reduces power costs.

Can some servers perform multiple roles?

Can you consolidate servers to do this?

Can some servers be powered down when they are not in use?

This thought process is the reverse of what you would normally think of doing, but that is what is required to be ruthless about cost reduction.

7.   Increase datacenter temperature to reduce cooling costs.

Every degree in the datacenter equates to a certain amount of power consumption to keep it that cool. If you know the maximum tolerances of your hardware temperatures, you can increase the temperature, thus reducing the amount of power used by the HVAC systems to keep it cool. This is a dangerous gamble because you can destroy your equipment if your calculations are wrong. A few degrees will not harm anything, and everything helps.

8. Remove all hardware and software lifecycles and focus on patching.

Plan on having your current hardware and software for the foreseeable future. Let your support contracts run out and support it internally only. Patch everything to the latest stable revision; firmware, software, everything. Do not run beta or RC code.

9. Sell all non-critical hardware on auction sites.

If you have enough inventory from your downsizing and decommission activities, then start selling it on auction sites like eBay, or privately via classifieds site like Craigslist. Take the revenue from these sales and put it back into the IT budget.

10. Partner with another organization to share office space.

If you are downsizing staff, then chances are that the office is starting to look a bit barren. I remember one company that I worked for had 10,000 square feet of office space for two people, including myself. Rent this extra space out to a company that can foot most of the bill.

11. Encourage staff brainstorming and participation in finding innovative ways to reduce operational expenses.

Gamify the process. Make it a challenge to continually cut costs and reward staff for their inventiveness. A little token can go a long way when showing appreciation.

12. Level with staff & ensure that they understand the state of affairs with no misconceptions.

Level set with the team. Let them know that they are performing a last stand and that it will require a different set of skills than regular IT jobs. It will require inventiveness, diligence and hard work. Every decision is an important one and though the pay may not be great, the rewards are great in terms of experience and IT street cred. Instill a sense of ownership and pride in the infrastructure.

13. Provide a flexible work policy with high-collaboration.

Encourage working from home and allow staff to set their own hours. Establish a results oriented work environment with accountability. However, also have constant communication through instant messaging, email, internal forums, web-meetings, etc. This will increase cohesion and reduce power costs in the office.

14. Make sure that team feels very close knit.

There should be a feeling of equality within the team. Everybody is doing their part and everybody feels valued. There is no room for any feelings of being slighted or infighting. You can think of it as a tactical team that supports each other and enables success. The confidence must be there so that there are no doubts in the abilities of each member, the decisions they make, or the method of execution of those decisions.

15. Make use of open-source monitoring and analytics.

Using monitoring and performing log analytics will reduce the mean time to repair (MTTR) and help the team be proactive in seeing potential issues. If there is an existing commercial monitoring solution in place, by all means use it to its full capabilities. If there is no monitoring solution in place, or if there are certain features that are not available, then there are many open source solutions that can provide these analytics.

If you look through all the suggestions above, you will see that many of them can be employed proactively or tested out in a bubble environment. The company does not have to be in its death throes before this list is looked at. It is important to understand the reasoning, motivation and trade-offs of each decision. What may make sense to a company with no resources, money or staff, may not be a good idea for a company that is just having a bad quarter.

8.10 Chapter Summary

1. It is possible to model comprehensive risk scenarios by using techniques that account for many variables and considerations.
2. Key Risk Indicators (KRIs) need to be monitored to determine the change in criticality of a risk.
3. The process of modeling of future states can augmented and the detail enhanced, by using the seven techniques named here:

   a) Pattern recognition
   b) Measuring trajectory
   c) Understand cycles and trending
   d) Interpret heuristics
   e) See the next state (Tetris block)
   f) Collect radicals
   g) Devil's advocate

4. The possibilities of future states and risk scenarios can be reduced by following the tenets of these philosophical razors:

   a) Occam's razor: The simplest explanation is often the right one.
   b) Halon's razor: Don't assume malice if something can be explained by stupidity.
   c) Alder's razor: If something cannot be determined by experiment or observation, then it is not worth discussing.

5. Internal politics within an organization can direct IT as much, if not more than any other constraint. It is important to understand the internal operations of an organization from that angle.
6. The market can create constraints within an organization that affect how an organization operates and by proxy, how IT must respond to support it.
7. Financial directives can change the mandates that IT adhere to, beyond just budget constraints.
8. When modeling risk scenarios, take into account regional environment considerations.
9. By using several techniques proactively and during the decline of an organization, you can extend the technological life of an organization.

8.11 Chapter Review Questions

1. Create a risk scenario for your environment based on a regional environmental event (i.e.: hurricane, earthquake, flood, etc.)

   a) Outline the scenario.
   b) List all systems and their dependencies that are affected.
   c) Define the impact on each affected system.
   d) List all staff resources affected.
   e) Define the impact to each staff resource.
   f) Define the impact to the business based on all the above factors.

2. Define Key Risk Indicators that should be monitored for a regional environmental event.
3. Outline a "radical event" (as outlined in section 8.3) and model several future states.

CHAPTER 9

Visual models

A good puzzle, it's a fair thing. Nobody is lying. It's very
clear, and the problem depends just on you.
--Erno Rubik (Creator of the Rubik's Cube)

---

*Spatial intelligence incorporates abstract and analytical abilities that go beyond merely seeing
images. Recognizing the image, knowing its relationship to other surrounding objects and
displaying the organizational structure of a thought are the key uses of visual models. This
chapter allows for a different perspective on risk criticality, dependencies and analytics.*

9.1 The Mind Map Model

Understanding and accepting possibility is a tricky business that is skewed by the realm of bias. Bias can be useful to focus perception, but it can also be wrong. Let's say that you wake up in the morning with no bias, no memory and everything is possible. How do you determine how to interact with the world? What comes next? Are you an ostrich? Are you a lemon tree? Are you a person? If you are a person, which person are you? What is your name and how has "this person" reached a point in their lives where they question what they are? What is the history leading up to this point in time?

Most people don't have existential problems that wake them up in the morning. That would be exhausting, somewhat traumatic and cause great difficulty to get out of the house on time. Imagine saying, "Sorry I'm late for the meeting, I thought I was someone else this morning, but it turns out I wasn't. Go Figure."

It is because of bias that we can accept certain assumptions without question and move on to other aspects of our day. Every single morning, we don't have to think about the question, "What / who am I and how did I get to this point in time?". Bias narrows our vision and gives us focus so we have fewer decisions to make and can operate efficiently on those assumptions.

Bias is a mental limiter that can provide benefits as well as hindrances. Think about the idea of a flat earth, or that the sun revolves around the earth, or that bloodletting can cure all ailments. These are flawed assumptions that have been debunked by people that were brave enough to contest the popular bias and beliefs of the time, and sometimes doing so at great personal risk!

So, what biases do you have that are wrong? What assumptions do you have about your infrastructure, your resources, capabilities, recoverability, etc. that are putting your environment at risk? Do you think that your core systems are so robust that there is no need for the expense of security? Do you think that obscurity equals security? Do you think that an old decommissioned Sun Microsystems server in the datacenter is a good place to keep your beer chilled? These are all flawed ideas that can be torn down by logic and by simply taking the time to consider it.

A mind map is a method of brainstorming that quickly lays out a train of thought and the associations between ideas and concepts. This can be used to map out your entire infrastructure, laying it all out on the table, in order to drill down into every area and comprehensively outline all possible, improbable and impossible risks.

How to get started:

1. Remove your assumptions and be aware of your biases and how they affect perspective.
2. Draw all your core services and systems in bubbles.
3. Think of every method of taking down the system. Brainstorm and link to the bubbles.
4. Once all possible ideas are exhausted, then go to the improbable, then impossible ideas.
5. Reference historical CVEs for the services or systems identified.
6. Find the weakest link.
7. For core systems map out undesirable outcomes and create a fault tree analysis model.
8. Create a spreadsheet and list out the core services, failure events and root causes. Give each item an ID. F# for core service failure, F#-E# for failure event and F#-E#-RC# for the root cause.

In the image below, you will see three levels of concentric circles with the core service at the center. In this example, we have:

Level 1 - DNS as the core service.

Level 2 - Possible failure events.

Level 3 - Possible root causes for each failure event.

Since this is theoretical, there are multiple root causes for each failure event.

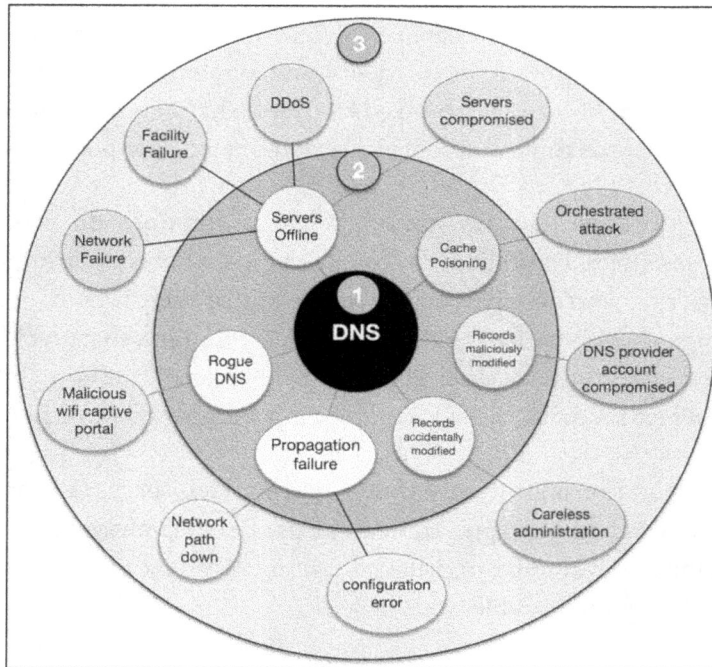

Figure 9- 1. Mind map for determining possible core service failure events and their root causes

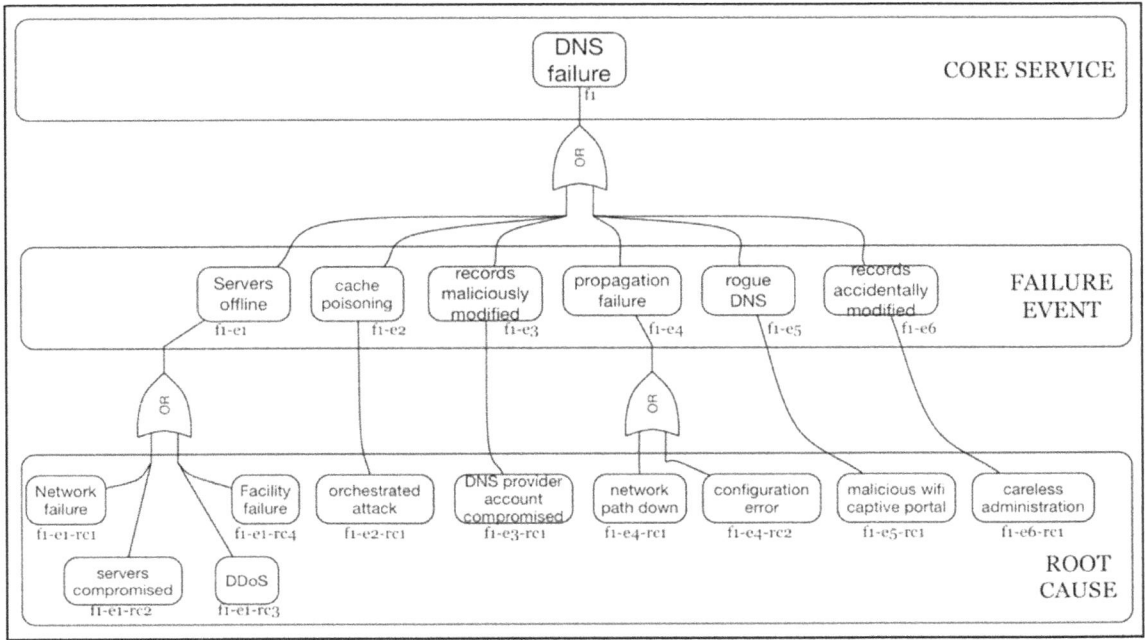

Figure 9- 2. Fault tree analysis for core services

Core Service ID	Core Service Name	Failure Event ID	Failure Event Name	Root Cause ID	Root Cause Name
F1	DNS	F1-E1	Servers Offline	F1-E1-RC1	Network Failure
F1	DNS	F1-E1	Servers Offline	F1-E1-RC2	Servers Compromised
F1	DNS	F1-E1	Servers Offline	F1-E1-RC3	DDoS
F1	DNS	F1-E1	Servers Offline	F1-E1-RC4	Facility Failure
F1	DNS	F1-E2	Cache Poisoning	F1-E2-RC1	Orchestrated Attack
F1	DNS	F1-E3	Records Maliciously Modified	F1-E3-RC1	DNS provider account compromised
F1	DNS	F1-E4	Propagation Failure	F1-E4-RC1	Network Path Down
F1	DNS	F1-E4	Propagation Failure	F1-E4-RC2	Configuration Error
F1	DNS	F1-E5	Rogue DNS	F1-E5-RC1	Malicious Wi-Fi AP
F1	DNS	F1-E6	Records Accidentally Modified	F1-E6-RC1	Careless Administration

Table of services, events and root causes

By going through this process, you start with the concentric circle thought bubbles for brainstorming. Then go to the failure analysis model which allows for more insight and Boolean logic (AND/OR gates), such as when multiple root causes are required for a failure event to occur. Finally, you end up with a spreadsheet that details all the underlying causes of the failure events. With the spreadsheet, you can create a to-do list of how to protect against it.

9.2 Heatmap Treemaps

A heatmap is used to determine a value from low (green) to high (red). It abstracts the numerical values to a spectrum that provides an easily understandable visual feedback. You may remember the "heat-vision" view from the Arnold Schwarzenegger Predator movie which created a heatmap based on temperatures. FLIR makes some great thermal cameras that you can use in the datacenter to spot cooling issues (environmental / facility risks). However, I'm getting off topic, so using a Predator movie reference, I'll "get back to the choppa."

A treemap is a method of presenting information that has nested rectangles. The size of the area of the rectangle represents a single dimension of data. The nested rectangles represent branches of the larger rectangles.

Combining both heatmaps and treemaps allow multiple dimensions of data to be presented in a quick and insightful manner.

For our purposes, we will have a series in the treemap for the core service DNS (black rectangle). The subgroups will be the failure events (grey rectangles). Then the root causes will be the rectangles with the color spectrum.

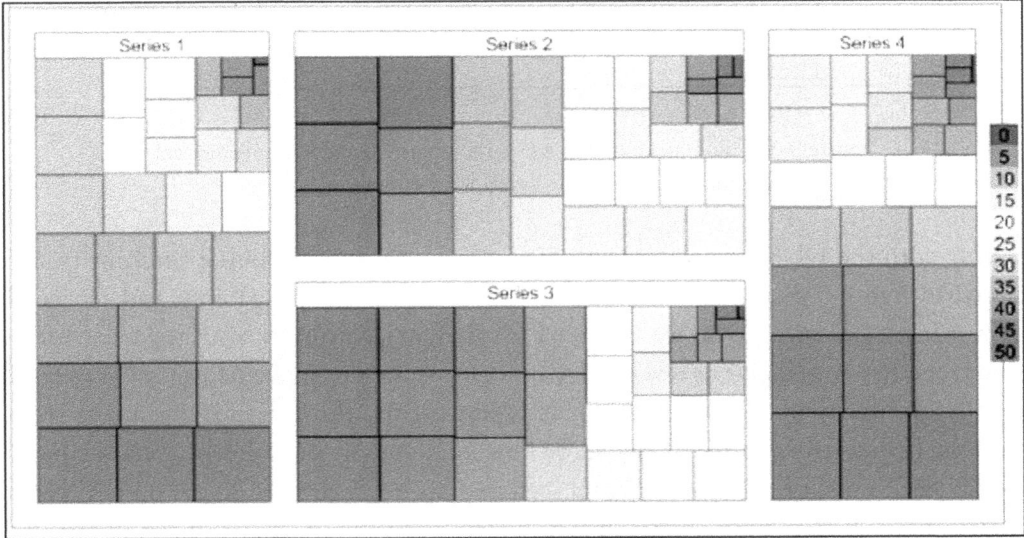

Figure 9-4. Example heatmap treemap model with multiple series

The probability determines the size of the area of a root cause rectangle it will occur. The range can be any scale you choose, but 1-10 is common. If we look at the area size regarding shape, we end up with the following shapes in figure 9-5.

It may look like a game of Tetris, but it is the easiest way to represent area in a rectangular form for whole numbers from 1-10. After 10, it gets difficult because to maintain rectangular shapes the numbers must be divisible by whole numbers. Prime numbers cannot be divided into whole numbers, so they end up being a straight line. Long straight lines do not work very well in a treemap.

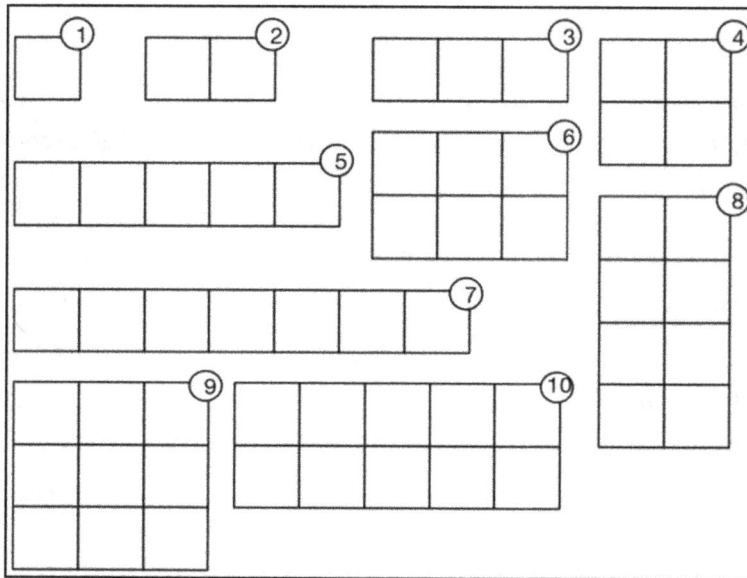

Figure 9-5. Calculating the area of a rectangle by using the numerical value for probability (or a very hard crossword puzzle).

Continuing with the DNS service example, we will add on two additional columns to the table we showed in figure 9-3. Those will have a numerical value for probability and recoverability. Normally when calculating risk, we will have a value for probability and impact, which when multiplied together, would give us severity. In this case, we are not concerned about impact because that has already been defined as the same for all failure events. The impact is the outage of the core service; thus, the severity can be calculated by how long the service is down. It's hard to estimate how long a service will be down if the exact scenario has not happened in the past. What can be estimated is the amount of time it would take to resolve a specific issue given the resources you have and the processes you follow. The ease of the process above is the recoverability, which we will use a 4-point scale to measure. The scale below describes this numerically as well as the associated color.

1 = easy (green)

2 = more difficult than not (yellow)

3 = hard (orange)

4 = difficult, or: buckle-up, you're in for a ride and it won't be fun (red)

These are the colors that will make up the heatmap for the treemap. Let's have a look at the updated table that shows services, events and root causes now that it includes probability and recoverability.

Recoverability is a subjective metric depending on who is defining its value. Management may have an idea of what it "should" be, based on RPO/RTO values and SLAs, but the operational teams will have a better idea of what it is. The best way to get an accurate estimate of the time it will take to recover, is to run simulations, or mock failures.

When conducting disaster recovery tests, scenarios are rarely used. The tests are usually centered around whether the services can be brought up within a stipulated timeline, optimizing the resources and workflows required to do so. I would suggest that as part of your next DR review and tests, that some time should be set aside for testing the response to each one of the root causes and failure events you have identified during this process. Military, police and emergency response units do this all the time to make sure that staff can respond in the best manner possible for the situation at hand. It's a core part of continual training and ensuring that everyone is on their A-game.

Core Service ID	Core Service Name	Failure Event ID	Failure Event Name	Root Cause ID	Root Cause Name	Probability (area 1-10)	Recoverability (1-4 & color green-red)
F1	DNS	F1-E1	Servers Offline	F1-E1-RC1	Network Failure	5	3
F1	DNS	F1-E1	Servers Offline	F1-E1-RC2	Servers Compromised	7	4
F1	DNS	F1-E1	Servers Offline	F1-E1-RC3	DDoS	7	2
F1	DNS	F1-E1	Servers Offline	F1-E1-RC4	Facility Failure	5	4
F1	DNS	F1-E2	Cache Poisoning	F1-E2-RC1	Orchestrated Attack	6	4
F1	DNS	F1-E3	Records Maliciously Modified	F1-E3-RC1	DNS provider account compromised	5	4
F1	DNS	F1-E4	Propagation Failure	F1-E4-RC1	Network Path Down	7	3
F1	DNS	F1-E4	Propagation Failure	F1-E4-RC2	Configuration Error	8	1
F1	DNS	F1-E5	Rogue DNS	F1-E5-RC1	Malicious WiFi AP	6	1
F1	DNS	F1-E6	Records Accidentally Modified	F1-E6-RC1	Careless Administration	8	3

Table of services, events and root causes with probability and recoverability

Now we can take our table and create shapes based on the probability number and the color based on the recoverability number. Using our DNS service data, we end up with the shapes in figure 9-8. The root cause ID has been put beside each shape and can be used in the treemap to identify it and trace it back.

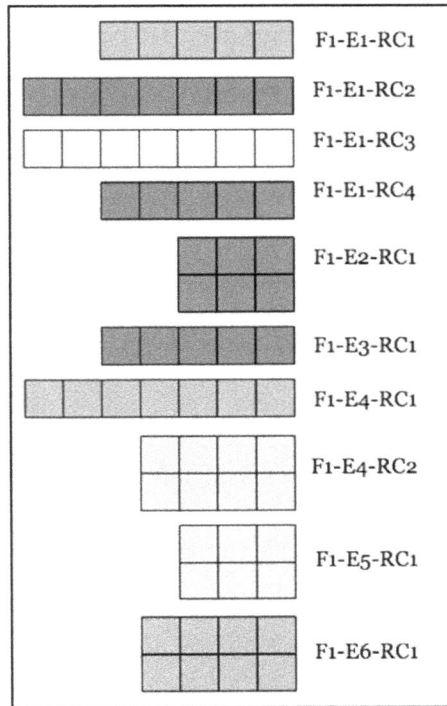

Figure 9- 6. Shapes used for Tetrimap

The less text in a treemap, the better. Since it is a visual model, text takes away from its impact. If you create a treemap that doesn't have any text, then each shape should be a hyperlink to the root cause ID.

Let's first have a look at the treemap created using the shapes we have here, then we will consider how it would look in a more traditional treemap. We use these shapes because they are easy to recreate without specialized software and can be done in Microsoft Excel, Visio, OmniGroup's OmniGraffle, iWork Numbers, or OpenOffice. If you have Microsoft Excel 2016 or newer, there is a treemap chart option built into it

This treemap, as I have said before is a bit like a game of Tetris. Therefore, I will call it a Tetris-Treemap, or Tetrimap (pron. tet-tree-map) for short.

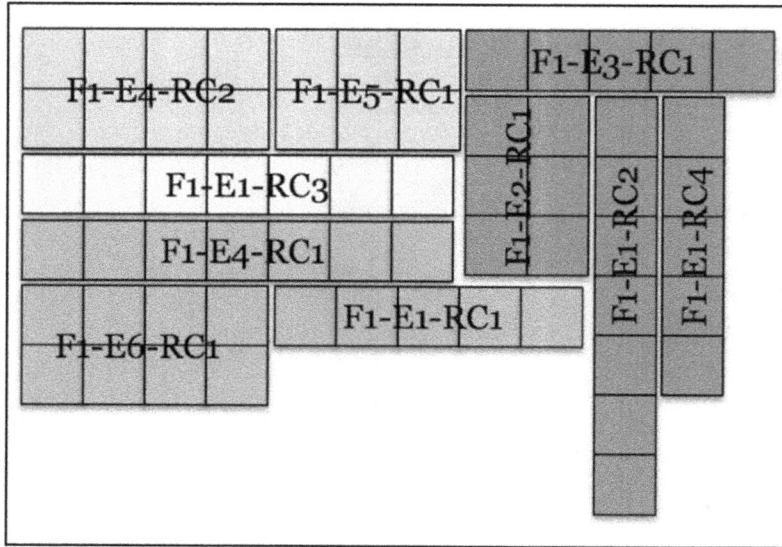

Figure 9- 7. An example of a Tetrimap

The Tetrimap is useful for quickly visualizing a heatmap treemap. However, it's a bit of an eyesore and difficult to read the root cause IDs because of the lines between the individual boxes.

A simplified Tetrimap (see figure 9-10) could look like figure 9-9, with the separating lines removed.

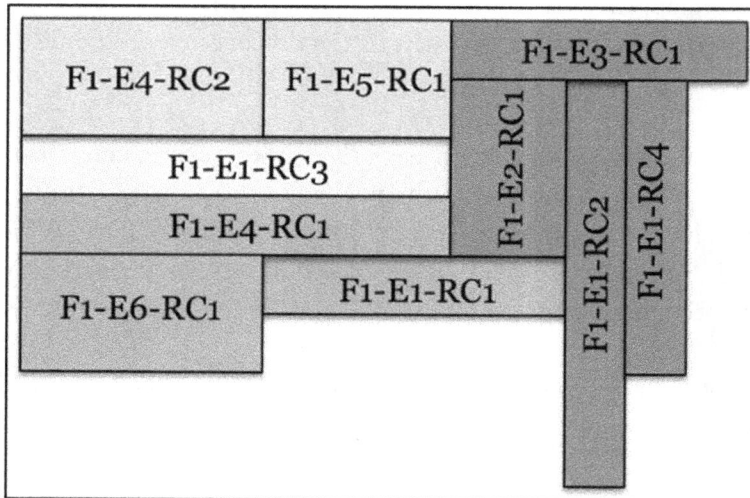

Figure 9-8. A simplified Tetrimap

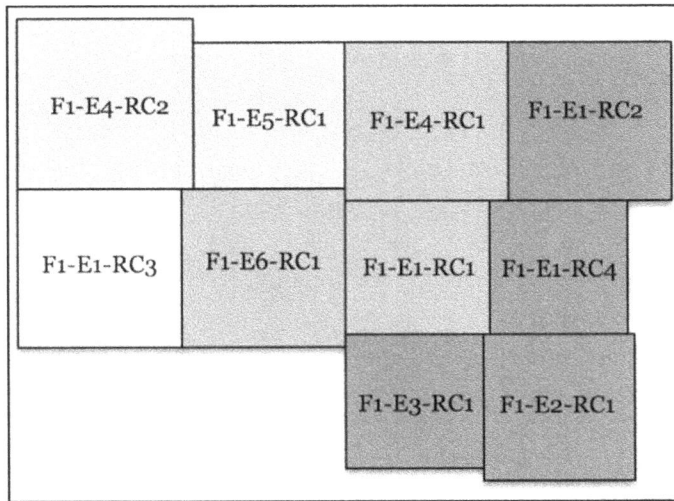

Figure 9- 9. A traditional treemap

A quick visual inspection of red areas shows a high probability for

F1-E1-RC2. This root cause is "servers compromised". This means that the environment is poorly equipped to deal with the fallout of a hack.

On the orange side, F1-E6-RC1 is the largest in the group. This root cause is "careless administration". This means that change management systems, standardized processes and role based access control may need some rework. Or perhaps there is a technical prowess deficiency in the pool of staff within the organization.

The heatmap gives insight into which areas or problems to tackle first to protect the organization. It allows you to create a list of prioritized action items for defense remediation. The goal is to have all small green squares if you cannot remove the risk entirely.

A more traditional view of a treemap will use square-like shapes as in figure 9-11. To achieve this, you need to determine the probability value (1-10) and then create a square with the length of the sides equaling the square root. For example, a probability of 7 would be a 2.646 x 2.646 sized square and a probability of 9 would be a 3 x 3 sized square.

One of the great things about treemaps is that they can provide a quick visual comparison of data that's orders of magnitude in difference of scale. However, with a traditional treemap a scale of 1-10 is not very different, so it is harder to differentiate in size. A Tetrimap does not have this same problem because the different shapes are similar enough to stack tightly together, but different enough to not mistake one value for another. This is another reason why I like the Tetrimap.

9.3 The Risk Register

A risk register is used to keep an inventory of every possible risk that may occur within an organization. It would have the following characteristics:

1. A risk ID associated with every risk documented.
2. A description of the risk.
3. A description of the impact.
4. A scale of the probability.
5. A scale of the impact.
6. A scale of the severity or criticality.
7. Recommended remediation or mitigation steps (also referred to as guidance).

Some other areas that may also be there, depending on the complexity of the register:

8. Key risk indicators (KRIs).
9. Residual Risk.
10. Whether it is a functional or non-functional risk.
11. The source of the risk (i.e.: the governance control).
12. Reference number from the source.
13. Audit testing method.
14. Artifacts (evidence that the risk has been verified).
15. Whether the risk complies or not.
16. Explanation of why it is not in compliance.

The difference between risk registers that have only the first seven items and those that have the sixteen listed, is related to complexity and how quickly a response action is needed. Using just the first seven, you can put together a working risk register in just a few weeks, create actions and start fixing identified risks incrementally.

If you have a very complex risk register, then it is being setup for the long haul. This is more about auditability and compliance, than it is about immediate action. This process may take months or years to complete.

Depending on the size of the risk register, it may be managed as a spreadsheet or in a database. I recommend starting with a spreadsheet to get started, populate it with as many fields as you need for providing an immediate result, then expand over time. When you get to the point where you have thousands of risks identified, then you may import the data into a database. However, if you have many people working on editing and adding data, then a database may be the better option from the get-go.

The ultimate risk register, in my mind, would be one that is immediately auditable, and tied to a ticketing system. If the audit testing method can be automated, then it can be performed on a scheduled basis as well as ad-hoc. This would let you know if an environment has risks that have cropped up from a new build, a change, or something breaking.

Some tools are available to perform a subset of these type of scanning and reporting activities, such as Microsoft's Security Baseline Analyzer, or Nessus from Tenable.

These tools are very beneficial because they allow for the audit of many risks, but they should not be used in the place of a risk register. A risk register is used to maintain a centralized point of reference for all risks, including both functional and non-functional. No scanning tool can address all risks.

One trap that some organizations fall into when they use risk registers is the illusion of control. This is when, even though they are following the methodology and process, they are bitten by a risk that causes a significant loss for them. This can happen for several reasons, but it often occurs because the probability or impact was incorrectly calculated.

I guarantee you that a risk register will provide value to organizations of every size and complexity. A risk register is not meant to be the pinnacle of all IT aspiration. It is a tool and strategy for providing incremental improvements, knowing where the "bodies are buried" and as a means of providing actions and direction for people to fix the issues.

Figure 9-10, is a sample of a simplified risk register:

Infrastructure Findings							Severity Index	very high = 85-100
								high = 75-84
								medium = 50-74
								low = <50
Category	RISK ID	Gap	Recommendation	Risk Description	Probability	Impact	Severity (probability x Impact)	
Hardware	R001	The hardware that is being used for the server DC01 is an HP DL380 G6. The warranty expired for this server in August 2012	Purchase a new replacement server and decommission the current one	A failure of the hardware without warranty will cause unplanned capital expenses and lead time for acquisition of new equipment. This will cause an extended outage.	8	8	64	
Hardware	R002	End of service life (EOSL) was April 2016. Although there are no current hardware alerts, this server has surpassed its recommended lifespan	Ensure warranty and support is available for the lifespan of the hardware	A failure of the hardware without warranty will cause unplanned capital expenses and lead time for acquisition of new equipment. This will cause an extended outage.	10	5	50	
Software	R003	There are mainly supporting applications on the server from HP, such as array management and monitoring. 3rd party software is installed from APC for UPS management and from Ubiquiti for Wifi management. Java is installed to run the APC and HP utilities.	Run servers as VMs to isolate host management from core infrastructure services	By not isolating management utilities, tools, third party software, etc. from core systems, there is the risk that a failure or issue caused by these will affect critical systems.	6	5	30	
Software	R004	System Updates were last performed two months ago	Perform Operating system updates / patches at least once a month	Microsoft releases major patches every month on the 2nd and 4th tuesdays of the month. By not patching immediately after the release, the systems are vulnerable and exploitable. This is excacerbated by having a server with sensitive information accessable with open ports on the internet.	6	8	56	

Figure 9- 10. Simplified risk register

9.4 Core Service and Infrastructure Components Selection and Justification

A core service or infrastructure component is something that is defined as essential to the functioning of the infrastructure or critical to the operation of the business. The following is a list of several core services and core infrastructure components with justifications for why they are classified that way.

9.4.1 DNS

DNS is as old as the internet, if not older. Most of what allows communication between devices can be boiled down to TCP/IP and DNS. If you want a device on one network to communicate with a device on a different network, then you either need to know its publicly addressable IP, or it's FQDN (fully qualified domain name). If you want to send an email to a user on a domain, then the domain needs to have an MX record. Here is a quick summary of some different types of DNS records:

A	Address Mapping records
CNAME	Canonical Name records
PTR	Reverse-lookup Pointer records
MX	Mail Exchange records
NS	Name Server records
TXT	Arbitrary non-formatted text string

If a DNS server is compromised, then that can invalidate all confidence in communication for devices using that DNS server. Websites, email servers and just about every service can be spoofed by a malicious actor. Or if a DNS server is the target of a DoS attack, then all devices using that DNS server will have slowed name resolution and thus a reduced or impaired experience. If the name servers are authoritative for a particular domain, then when attacked, that domain may become inaccessible rendering all associated services down.

In 2016, a very large DNS provider (DynDNS) was attacked. This ended up taking down many major internet destinations for several days.

9.4.2 Network

The network is the fabric of communication. It is the routing of traffic from one device to another, specifying paths and determining the quickest and most efficient way from point A to B to C. A network outage can be caused by misconfiguration, attacks, or hardware and/or software failures.

Disrupting network services can cause outages for some users or devices, disruption of communication between dependent services (causing secondary service outages), the contention of resources when accessing service causing slowness and queueing of processes and secondary service outages when network communication to storage systems are impaired.

9.4.3 Storage

All applications, services and data need to reside somewhere. Even if they are running entirely in memory, they still need to retrieve data and store data from storage somewhere. When services are unable to access the dependent storage, they will halt, crash and burn.

9.4.4 Compute

If processes need to run (which they always do), then you need to run them on something. The more processes or instructions that need to run, the more compute is required. Runaway processes, memory leaks, bad coding, hardware failure, etc. can make compute resources unavailable. The fewer compute resources available the higher the contention for processor time. This causes wait times, which slows down all processes sharing the affected compute resources. Or if there are memory hogs that are unaccounted for, or uncontrolled, those could cause processes that run quickly to start running orders of magnitude slower. However, when this happens and you manage the compute, don't panic. The network team will always be blamed first.

9.4.5 Authentication, Authorization and Accounting (AAA)

How do you manage a system? You log into it as a user and perform actions that are allowed. What you do, how long and what is touched, can be logged. These are the elements that make up AAA.

If your authentication services are unavailable, then nobody can login as a named user. This means that permissions and capabilities are impaired. It is a pre-cursor for an awful scenario, where something is broken, but you cannot fix it, because you cannot login.

If your authorization parameters are incorrect or rather, in a non-desired state, then you cannot access the resources you require or want. This will impair you from achieving your desired objective. If certain permissions are required for a service or function to work, then it will become unavailable if the access is not granted.

If a tree falls in the forest, does it make a log? How do you know what has happened in your environment if there is no record of it? Accounting (or logging) is crucial for knowing who did what to something and when it occurred.

9.4.6 Environmental

Environmental factors encompass everything to do with the facilities and the specified location. There is a reason that organizations that have put some thought into their DR strategy have sites in very different geographical regions.

Some regions are prone to recurring issues, like flooding, ice storms, tornados, hurricanes and earthquakes.

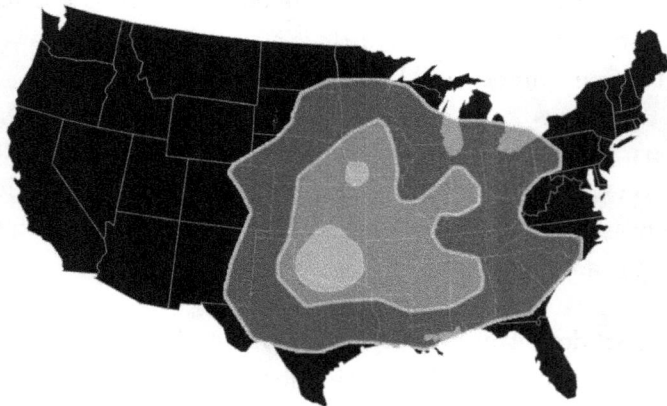

Tornado Risk — Lower / Higher

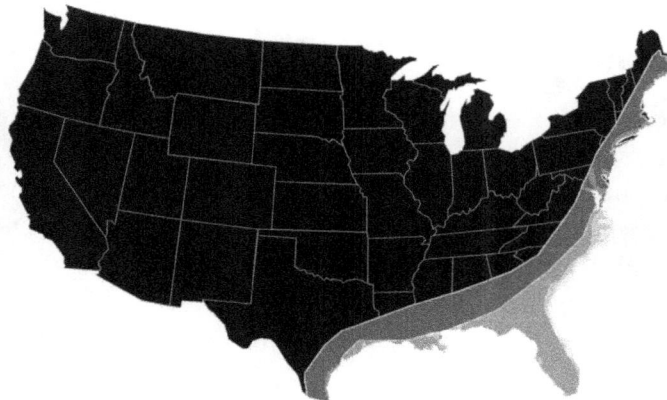

Hurricane Risk — Lower / Higher

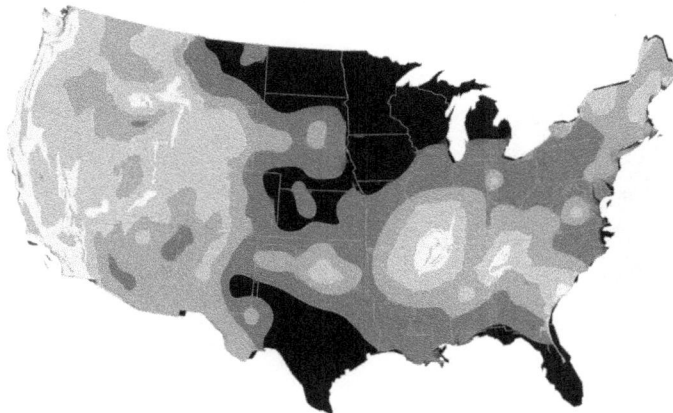

Earthquake Risk — Lower / Higher

Figure 9- 11. Natural Disaster regions in the continental US.

When thinking about where a data center should be located, it is a good idea to think about what environmental factors may affect that region. Looking at the map in figure 9-11 for the continental United States, you can take your pick of calamities based on the area.

The eastern seaboard gets hit with hurricanes, the central region gets tornados and the west coast gets earthquakes. The riskiest area (statistically speaking, due to environmental reasons) to have a data center located in is Dallas, Texas. It has recurring tornados, hurricanes, earthquakes, floods, hail storms and droughts. What happens when a natural disaster occurs? The first thing that typically occurs is that the mobility of staff is restricted. If people need to be in the data center to keep operations running, then they might not be able to physically get there. Power lines may go down, forcing the datacenter to run off a diesel generator. If access is restricted because of weather or roads being blocked, then the fuel tanks may not be able to be refilled. External network connectivity may go down if links are severed or line damage occurs. The facility itself may also have structural damage and internal systems may fail.

9.4.7 Power

Power can be complicated, or simple, depending on what your role within the organization is. Application owners do not care in the slightest about power, it's just an assumption that it will be available. Operations needs to make sure that the equipment in the datacenter is evenly distributed across PDU's (power distribution units, or another name for big expensive power bars for racks in the datacenter) and power channels. The assumption with operations is that there will be redundant power channels available backed by independent UPS circuits with clean, regulated power and that the UPS units are backed by generators. Datacenter operations/facilities, will be looking at the power utilization of each PDU, the power spikes from boot storms, the power usage efficiency (PUE) and the datacenter infrastructure efficiency (DCIE). It is the assumption of the datacenter that they get clean regulated power from the local utility company.

Any time there is an issue with any part of the power system, from anywhere between the utility company and a device, means that the applications and services running on the environment are at risk. Clean power means that there are no voltage spikes or brownouts from the utility company. Dirty power means that the variations that occur in the power distribution need to be compensated for by equipment in the datacenter. When this occurs, it will often put undue load on the UPS units and increase the chance for failures to occur. As failures occur, redundancy is adversely affected, which puts all applications and services hosted at that datacenter at risk.

9.5 Questions to Ask to Help Build the Risk Register

When building the register, it is important to get insight from multiple sources.

Some document sources are:

- Audit reports
- Technology plans
- DR testing and readiness reports
- Vulnerability assessments and
- Service ticket reports from the last year

The people that should be involved in the creation of a Risk Register are:

- Management
- Operations
- Representation from administration
- Human resources
- Any other discrete department that may have access to different systems or follow different processes

It is also good to get insight from customer sentiment, competitors and industry analysts. Being tactful is important when approaching these segments, because you do not want to get information that is biased, based on where the questions are coming from.

When speaking to stakeholders, the free-flow dialog that occurs can be described as a risk identification workshop. During this process, it is common to identify risks in the following categories:

1. Technical Risks
2. Integration Risks
3. Quality Risks
4. Strategic Risks
5. Management Risks
6. Planning Risks
7. Legal Risks

Due to the technical nature of this book, I will be focusing on the top three categories.

1. Is there a current architecture design document in place for all technologies being utilized?
2. Is there a change management system in place? Can changes in the past few years be audited?
3. Who is responsible for the overall technology plan?
4. Have there been any mergers, acquisitions or divesting of interests in the last few years?
5. Is there an existing vendor relationship in place for each particular technology area?
6. Do different departments have different technology acquisition processes?
7. What is the vetting process for technologies and solutions?
8. Is there a way to track deviation from the overall technology plan?
9. What are the top 10 support issues over the past year?
10. What are the most common failure areas?

These questions will start you down the path to get a dialog going, but they are far from comprehensive. I suggest at least 50 questions to start a risk identification workshop, then subsequently follow-ups to continue the dialog as more information is uncovered.

9.6 Chapter Summary

1. Use mind maps for brainstorming risks and causes.
2. Use fault trees to determine events that occur from a single root cause, or a combination of several root causes.
3. Use heatmap treemaps to gage your current level of risk and key areas for strategic resolution.
4. Create a risk register so that you have a record of all identified risks and can track them.
5. Map out risks to all your core services.
6. Talk to stakeholders in many different areas for a composite perspective when risk reporting.
7. Look to internal sources such as documentation, as well as external sources such as industry analysts and competitors for a comprehensive risk register.

9.7 Chapter Review Questions

1. What is the easiest method of disrupting business within your organization?
2. List the 3 most important technologies in your organization. Now remove them from the equation. Pretend that you can no longer use them. What would you do to get your business up and running again?
3. Think of the 5 most technologically influential people within your organization; the ones with the most responsibility. Now pretend that they all left the company at once with no notice. What would you do to get your business up and running again?

CHAPTER 10

Red Teams and Robots

"If they didn't want you to get inside, they ought to build it better."
--Harold Finch (Person of Interest TV Series)

How do you really know what your weaknesses are? What risks did you miss in your analysis? The only way to know for sure is to try to break it. This chapter is about how to find the skeletons in your infrastructure's closet.

10.1 What is a "Red Team"?

Red Teams are groups that are specifically tasked with finding weaknesses in an organization's strategy, technology, methods or processes. The idea is to provide a friendly simulation to frame the capabilities of malicious actors against the internal environment. However, before I dive into the specifics of how a Red Team operates, let's take a step back to understand, where the term originates and the significance of the color red.

Red and blue are some of the oldest dye colors that have been used by humans in clothing and textiles. However, dyes were not as cheap and prevalent as they are now. So historically, the brighter the color, the greater the display of wealth and power.

Red was used by the British for hundreds of years, which is why they were known as the redcoats. (However, in the late 1800's, the British military dropped the red for khaki and drab colors). Red is also associated with left-wing politics, such as with communism and socialism, as seen with China and the former Soviet Union utilizing bright red nationalistic colors.

During the "Red Scare" in the U.S. in the early 1900's, red flags were banned to prevent any rise of communism. The state of Oklahoma used to have a red flag with a white star. After the law had been enacted, it was changed to a blue flag.

Red is also seen as a color of defiance. When a red flag would be hoisted in battle, it meant that they would take no prisoners, meaning that it was a fight to the death, regardless if they are attacking or defending. This happened in the battle of the Alamo during the Texas Revolution in 1836, where all defending

> Texian, as opposed to Texan, refers to colonists living in the Republic of Texas before it became a U.S. State in 1845.

Texians were slaughtered by the Mexican militants while a red flag was raised.

Historically, blue has been used by the French, the Scottish and Americans. You can have a look at military uniforms during the Hundred Years War, the Napoleonic Wars and the American Revolution. The United Nations flag is also blue. Red versus blue has a very long history and still survives today in many regards. Sports such as Tae Kwon Do and Olympic freestyle wrestling require one combatant to wear red and the other to wear blue.

When western militaries conduct war game simulations and maneuvers, the opposing force is called the Red Team and the defending force is the Blue Team. Militaries were the first to make use of the concept of Red Teams, but as time went on large government contractors, or those with critical assets and information saw the value. Nowadays, many organizations make use of Red Teams to identify weaknesses and exploitable regions within their infrastructure.

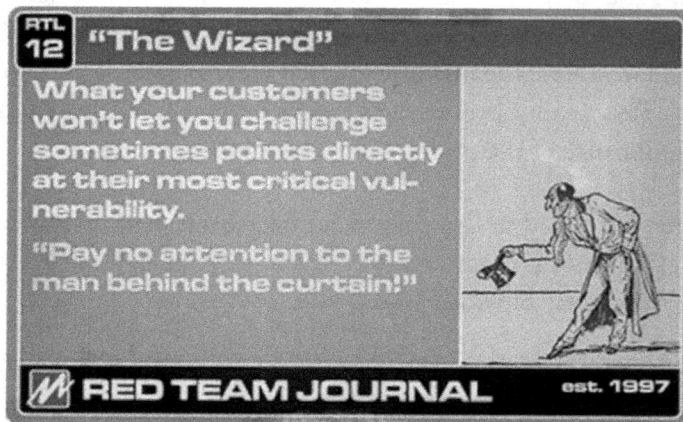

Figure 10.1 - Red Team Journal card

Red Teams are used to compromise live assets in a friendly manner. This may consist of infiltrating a location, gaining access to critical data, then exfiltration of that data. Or it may be the planning of an attack on a physical location, an organization, or the public. A Red Team is seen as an opposing force acting in aggression in a controlled environment.

This used to be only a physical endeavor, but it is frequently becoming a cyber endeavor as well. This is also the case with war-game scenarios. The reliance on technology is becoming a primary focus within war-games. This is not to say that the focus should only be on technology, though psychological warfare, or Psy-Ops, can also play a big part and have a big payoff in war-game scenarios. Misdirection, social engineering and de-motivational attacks can also compromise people that have access to the resources or desired assets.

Regardless of the attack vector, there is a large psychological component to a successful attack. In order to understand the thought process, there are some rules / guidelines that are used to direct the actions of the team. The Red Team Journal (http://redteamjournal.com/) is an excellent blog that promotes the practice of Red Teaming, alternative analysis, and wargaming. They have put together a list of "laws" that should be understood and heeded when Red Teaming. Here are a few below:

RTJ Red Teaming Law #8:

Risk is subjective. Oh, and goals are mercurial, perceptions are plastic, knowledge is gettable, time is exploitable ... Review this law whenever you think you've mastered the practice of Red Teaming.

RTJ Red Teaming Law #9:

Red Teaming is not forecasting; Red Teaming is the art of challenging assumptions and exploring the possible.... although your forecaster will always benefit from talking to your Red Teamer.

RTJ Red Teaming Law #15:

The apprentice Red Teamer thinks like the attacker. The journeyman Red Teamer thinks like the attacker and the defender. The master Red Teamer thinks about the attacker and defender thinking about each other. Hire an apprentice to model an unsophisticated adversary. Hire a journeyman to model a sophisticated adversary. Hire a master to model the system.

RTJ Red Teaming Law #22:

Unexpected surprise is what happens while you're waiting for the expected surprise. Think tanks and pundits specialize in expected surprise.

RTJ Red Teaming Law #34:

("Question"): In many ways, the art of Red Teaming is the art of asking the right questions, from the right perspective, at the right time. Ask the wrong questions, and it almost doesn't matter how well your Red Team performs.

RTJ Red Teaming Law #40:

("Der Zauberlehrling"): Take heed; we can build systems of systems of systems, but that doesn't mean we can always secure them."

10.2 Threat Profiles

When a Red Team performs their operations, they have the capability to act as aggressively as they warrant is necessary. Depending on the target and the desired effect, the threat profile is established and acted upon. Will it be quick and dirty? Or stealth-like and invisible? Is the objective data exfiltration? Or a Denial of Service attack? Is one attack used to mask another, or is it a probe to see how people respond? How many people (technical resources) are required and what is the duration of the attack?

All of the above questions are what is used in cyber adversary modeling to build a threat profile. This can be used to determine the details of the attacking parties (such as the Red Team). Once the threat profile for an attack is inferred, then risk response strategies can be put in place.

THREAT LEVEL	COMMITMENT			RESOURCES			
					KNOWLEDGE		
	INTENSITY	STEALTH	TIME	TECHNICAL PERSONNEL	CYBER	KINETIC	ACCESS
1	H	H	Years to Decades	Hundreds	H	H	H
2	H	H	Years to Decades	Tens of Tens	M	H	M
3	H	H	Months to Years	Tens of Tens	H	M	M
4	M	H	Weeks to Months	Tens	H	M	M
5	H	M	Weeks to Months	Tens	M	M	M
6	M	M	Weeks to Months	Ones	M	M	L
7	M	M	Months to Years	Tens	L	L	L
8	L	L	Days to Weeks	Ones	L	L	L

Figure 10.2 – Threat profile rating

In the threat profile, some levels dictate the possibility of success for an attacker. These range from 1-8, with 1 being the most possible and 8 being the least.

The objective for attackers with a threat profile of 8 may be the same as a threat profile of 1, however they would be achieved by very different means. It is difficult for any organization, regardless of whom they are, to protect against a level 1 threat. According to Verizon's annual Data Breach Investigation Report (DBIR), many organizations that are breached do not know that they have been until several months after it has occurred. By monitoring for indicators of compromise (IOC) and anomalies internally as well as externally, the damage caused by a breach can be minimized and counter measures can be employed.

The levels at which a threat is posed can be simplified by categorizing them as high, medium and low. A high-level threat is posed by an organization of highly technical personnel with considerable resources and time. This type of persistent attack must be highly funded, organized and determined. The attacks can last decades and can be considered a theatre of war. State sponsored offensive cyber-war organizations and insurgent forces come to mind.

A mid-level threat needs a well-organized attacking force with considerable technical skills that operate over an extended period. Advanced persistent threats (APTs) and coordinated attacks come to mind. These are typically operational and have a short-term objective. Mercenary hacking groups and specialized cyber tiger-teams fit this profile.

Resources that are not very skilled, organized or determined to execute a low-level threat. Opportunity crackers, script kiddies and cyber-criminals often populate this region. This does not mean they won't succeed, but rather that the likelihood of success is lower. Though if not adequately defended against, and given enough time, an organization can still be compromised by attackers of this profile.

10.3 The Difference Between Vulnerability Assessments, Penetration Testing and Red Teaming.

It is important to note the differences between vulnerability assessments, pen testing and Red Teaming as there are similarities and they might be used interchangeably by those unaware.

A vulnerability assessment (VA) is a scan of all assets to determine what patches/fixes are not current and can be exploited. Only known vulnerabilities are addressed and a report is created for remediation. The report then becomes actionable as operation teams patch all patchable assets. The assets that are not patchable, due to old versions of code that are no-longer supported, are entered in to the risk register. If the risk warrants a response beyond acceptance, then a strategy must be put in place to isolate or remove the vulnerable asset.

A penetration test (Pen test) is performed when all pertinent actions from a VA have been pursued. This is performed by one or more individuals that have been authorized to attack an environment (without causing operational outages) in order to meet an objective such as gaining elevated access rights or accessing sensitive data. Pen tests are not widely publicized to the entire organization, but it is not performed in an overly secretive manner. Marks of brute force attacks and lack of stealth are seen in these sorts of engagements. The deliverable from a Pen test is a report detailing what was performed, with step by step logs and timelines. An entire attack session may be recorded with screen recording software by the attacker, or it may be recorded with text log files and screenshots. The action that is required by operation teams after the fact, is to prevent the method of exploitation from occurring in the future. This may involve changing physical hardware topologies, software or processes.

Goal-based Pen testing has a lot of similarities with Red Teaming. Both use a set of tools to determine weaknesses in an infrastructure and then exploit them. However, the toolset and methods that a Red Team uses does not have to be pre-approved and the people who are "in the know" of the engagement are very limited. This ensures that it reflects the profile of an actual attack.

Some things may be pre-approved and others require more detailed authorization procedures. In 1993 Randal Schwartz, of Perl fame, complained to Intel IT that they had a poor security framework. The IT team didn't listen so he went ahead to prove his point by cracking passwords and gaining access to more sensitive systems. Although he was a contractor for Intel, he had no permission to do this. He proved his point but ended up as a criminal in the eyes of Intel legal team. He was charged in 1995 with 3 felony accounts of altering a computer system without authorization. In 2007, his criminal record was expunged and he was no longer considered a felon. It has been said that he was guiltier of bad judgment, than a criminal act.

The stealth of the attack is also more important for a Red Team because they must account for the possibility of detection. This will limit the methods of attack to ones that are less apparent. A Nessus scan (or other bulk interrogation method) for instance, may get flagged by monitoring systems and alert security operations (blue team). This requires the attacking toolset to be more innovative and use methods that are not just technical.

The most successful methods of gaining access to the internal network are via phone and email. Social engineering, phishing and whaling are all methods used to compromise people; the weakest security link. All the security controls in the world that apply to technology, will not stop a person from being compromised through social engineering. If those methods fail, then external network interrogation is performed.

If physical security is being evaluated or social and external attempts fail, then surveillance and research is performed before any attempt at access are performed. What are the patterns of movement and access? Where do employees park? Is there a perimeter security system? Are there dead zones (areas that are not monitored by the security system), or can the security system be evaded or fooled? What are the access protocols and procedures? Can keys be forged or proximity cards cloned? Can in-person social engineering be performed? How can human nature be exploited?

Penetration testing provides more control over the attacking process. The attacking and defending teams work closely and there are constant updates to the client of the progress of the testing team. A wider variety of attacks can be performed and the results evaluated in real time by both sides.

If an organization has performed enough iterations of vulnerability assessments, and it has reached a level of security maturity that it is comfortable with, then Red Teams can be brought in to give a bit of a real world "kick in the teeth", so to speak. Another reason for bringing in a Red Team is to test out incident response, the Security Operations Center, or a Security Information and Event Management system in a live scenario. When not given warning of an incident, can your staff follow procedure in a timely manner, or do they become easily flustered. Remember, time is of the essence in responding to security incidents. Often when companies perform recurring attacks, they will hire different Red Teams in order to gain a more diverse experience that reflects the real world. These are often performed on a recurring basis, whether quarterly or yearly.

Organizations that have an immature security practice will not get much benefit from Pen Testing or Red Teaming, because there are so many vulnerabilities organizations, the first step is performing vulnerability assessments, patching, exposed that it would be like a boxing match with a blind person. For those holes and creating processes and policies to strengthen their infrastructure.

The Center for Internet Security (CIS) has outlined the top 20 security controls that prepare an organization to have a security posture that is mature enough to benefit from a Red Team. https://www.cisecurity.org/controls/. The first five controls will eliminate the vast majority of vulnerabilities. The remaining fifteen controls will help secure the organization from the most pervasive threats that are seen in the wild today. Beyond these 20 controls, Red Teams can assist in uncovering less obvious vulnerabilities or those requiring more sophisticated attacks.

Here are the top 20 security controls:

1. Inventory of Authorized and Unauthorized Devices
2. Inventory of Authorized and Unauthorized Software
3. Secure Configurations for Hardware and Software
4. Continuous Vulnerability Assessment and Remediation
5. Controlled Use of Administrative Privileges
6. Maintenance, Monitoring, and Analysis of Audit Logs
7. Email and Web Browser Protections
8. Malware Defenses
9. Limitation and Control of Network Ports
10. Data Recovery Capability
11. Secure Configurations for Network Devices
12. Boundary Defense
13. Data Protection
14. Controlled Access Based on the Need to Know
15. Wireless Access Control
16. Account Monitoring and Control
17. Security Skills Assessment and Appropriate Training to Fill Gaps
18. Application Software Security
19. Incident Response and Management
20. Penetration Tests and Red Team Exercises

10.4 The Red Team Process

When a Red Team is engaged, the following process occurs:

1. The CISO engages 3rd party Red Team.

- This may occur if the organization wants to test its security maturity in a real-world scenario, or if the CISO doesn't have buy-in for expanding the security budget and needs to prove the requirement.
- In some very mature organizations even the CISO is left in the dark. That way they can test the response up to the remainder of the C suite.

2. Red Team is just given a company name.

- They are not given any special access to the company or its information. It is part of their charter to find out whatever information they need to perform the attack.
- Research is done on the organization, hierarchy, locations, employees, and social engineering strategies that may be applicable.

> 'Pen Testers and Red Teamers are in high-demand, arrogant, adrenaline junkie, high-maintenance geniuses'
>
> -Red Team Manager

3. Phone campaign is used to ferret out more information and set stage for email campaign.

4. Email phishing and whaling campaign initiated.

4B. If the campaign was not successful, then external attack is required.

5. If the campaign is successful, then network topology is explored and stealthily mapped.

- Servers, routers and other devices are mapped out and identified.
- Vulnerabilities are found for specific versions or code running on devices.

6. Exploits are run to gain access to devices and account escalation techniques are used to get administrator level control.

7. Critical data is exfiltrated and the process is recorded with screen captures and detailed application logs.

During the process, the team communicates via email when key events (or flags) are achieved. The client is only engaged in the process if they need to formulate an immediate plan to address an immediate critical threat. An example of a deliverable would be an executive summary and a technical report. The findings would be sent to the client and then a couple of days later, a presentation to the CISO and the organization's blue team (security division) is performed. The de-brief is often a verbal recap and a video demonstration of the exploitation and penetration process. The project is then closed out and remediation actions for improving security against the attack methods and vulnerabilities are outlined.

The ability to manage a Red Team takes a lot of effort and expertise in several different disciplines. There is the interaction and planning with C-Level execs, technical teams (blue and red), project management, a deep security background and technical resource management. Red Teamers and Pen Testers are not your average technical resources either. A special ability to engage, retain talent and get results is required to manage them. Recently, I conversed with a Red Team manager and they put it like this: "Pen Testers and Red Teamers are in high-demand, arrogant, adrenaline junkie, high maintenance geniuses."

The industry is gaining more acceptance of the practice of introspective security testing like Red Teaming. White hat or grey hat hackers attack networks to make them stronger are becoming in greater demand as it evolves into a common practice area of security operations.

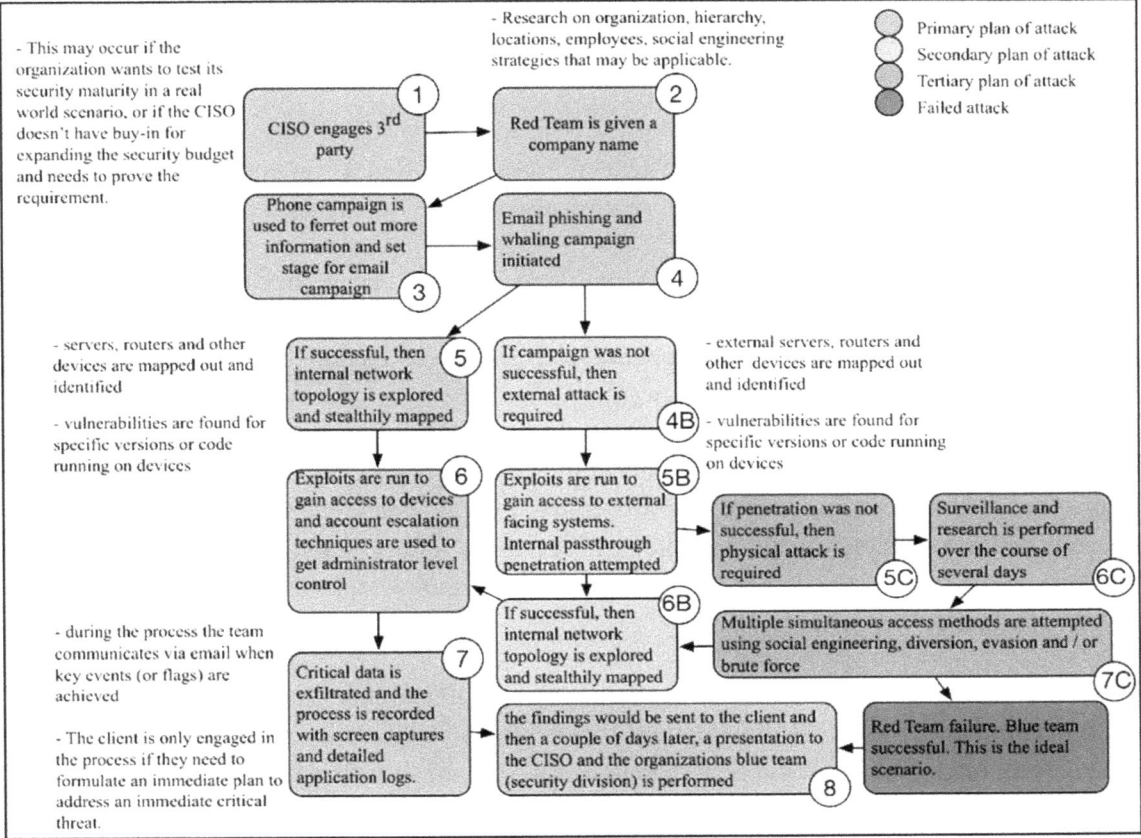

Figure 10.3 – Red Team process

10.5 A Look Inside Red Teaming

In the previous sections, we covered the general process of a Red Team engagement. Now we'll look at the thought process and methods that are used by the Red Team themselves.

When a Red Team engagement scope is being defined with a client, they need to know a little about many aspects of the organization. They will need to know about the organizational structure, internal politics, business needs, risk appetite, etc. They need to have a response to objections prepared before discussions with clients occur.

There are two main categories of Red Teams in the enterprise space: software development / code security, and infrastructure security. The former will look at reverse engineering, finding holes in the code design and exploiting vulnerabilities in specific programming languages and application frameworks. The latter looks at the larger infrastructure topologies, finds the weakest links, then exploits them.

Both will make use of threat analysis techniques and vulnerability scanning. Red Teams may perform engagements with clients before software development lifecycle release, before deployment of new technologies or infrastructure changes or as a means to validate a mature and hardened environment.

Some other things that are performed by the team independently of an engagement are the development of "Proof of Concept" (POC) exploits for vulnerabilities and just generally staying on top of vulnerabilities affecting in use operating systems and applications. The landscape is constantly evolving on a daily basis and it would be nearly impossible for the internal security staff of an enterprise to have the same level of skill and exposure that a professional Red Team has.

Red Teams are also often brought in to the following engagement scenarios:

- Compliance testing requires that an impartial 3rd party perform the scanning.
- On key assets, before firewall changes are approved.
- On software and systems that are running new code. Before they are put in to production or before the code is released publicly (in the case of a software company).
- On assets before DNS changes are approved.
- On templated / standardized OS builds.
- On all new systems before they go live.

When an attack is in flight and the Red Team is at the point where they have access to hosts, this is the methodology they use:

1. Assess threat vectors.
2. List and scan hosts.
3. Look at previous scan results from others for similarities.
4. Review findings and perform deeper analysis.
5. Apply CVSS3 scoring.
6. Internal discussion / dialog.
7. Perform pen tests.
8. Review findings and test deeper for several iterations.
9. Document and log findings for presentation to stakeholders at end of engagement.

Red Teams are often small teams that have several pen-testers with subject matter experts in specific areas. There are also vulnerability analysts, project managers, a team and a department lead. Continual education is viewed as a part of the job and is required and scheduled for each member of the team.

Keeping on top of all possible exploits, and how they look in the real world, means that a proof of concept for a specific exploit needs to be performed. This may entail programming a tool or a workflow to compromise the target based on the loose details of the exploit method or process.

One of the problems that people in Information Security find, is that they are often butting heads directly with the old establishment. These are the people that have forever done things one way and don't like it changed on them. Security does not often work as an enabling force, but rather a restrictive one. If security is not built into the culture or DNA of an organization, then it is often vilified or circumvented by staff, because it is viewed as a roadblock. Other teams may even say that "security puts the NO in inNOvation". The security budget is viewed as a "nice to have", but not critical for the business to make money.

For security departments, a way to combat this is by relating the need for security to a cost associated with risk. The justification can be bolstered by the deliverables of a Red Team. This way, the risk criticality can be supported with real proof. Then, required resources can be brought to bear on the vulnerabilities within the organization, and even assist with justification for new initiatives like software defined networking.

An infrastructure (or sub-portion thereof like an application stack) will be in one of 4 states:

1. Unpatched / legacy

These environments are usually just spun up and then forgotten like a dinner party where people left the food out all night; sometimes it's salvageable with updates, but you might find some bad bits that make you reconsider the effort required. In those cases, double bag it with a micro segmentation solution.

2. Managed

These environments are the ideal state, fully managed and maintained.

3. In-Flight

These environments have not been deployed yet. They may be in or out of compliance standards, but they probably have not gone through the rigor of the patching and hardening teams.

4. Road mapped

These are environments that are being planned for the future.

Environments that are in state 1 (unpatched / legacy) should be either phased out or micro segmented, as they are a huge security risk.

Environments that are in state 3 or 4 can be easily converted to a fully managed state, but security must be integral to the design, as opposed to be bolted on or reworked afterwards.

Red Teams can uncover the holes in states 1, 2, and 3 and they can be used at each stage of an infrastructure's lifecycle. However, stage 1 would be better served by a comprehensive and iterative vulnerability and patch engagement beforehand.

The first battle will be won by justifying value of the team. A strong initial deliverable will do wonders for expressing the necessity of Red Teaming to the organization and management specifically. Quantifying risks regarding monetary and intellectual property loss and the probability of those risks is key in creating one's argument for Red Teams.

10.6 Chaos Engineering

In Japanese, there is a word called "Ma" that translates into the "gap", or "space" between things. It is more than that though; it is the experience of pause, interval, and spatial awareness. It is experienced in the mind and exists because it is observed in balance or lack thereof.

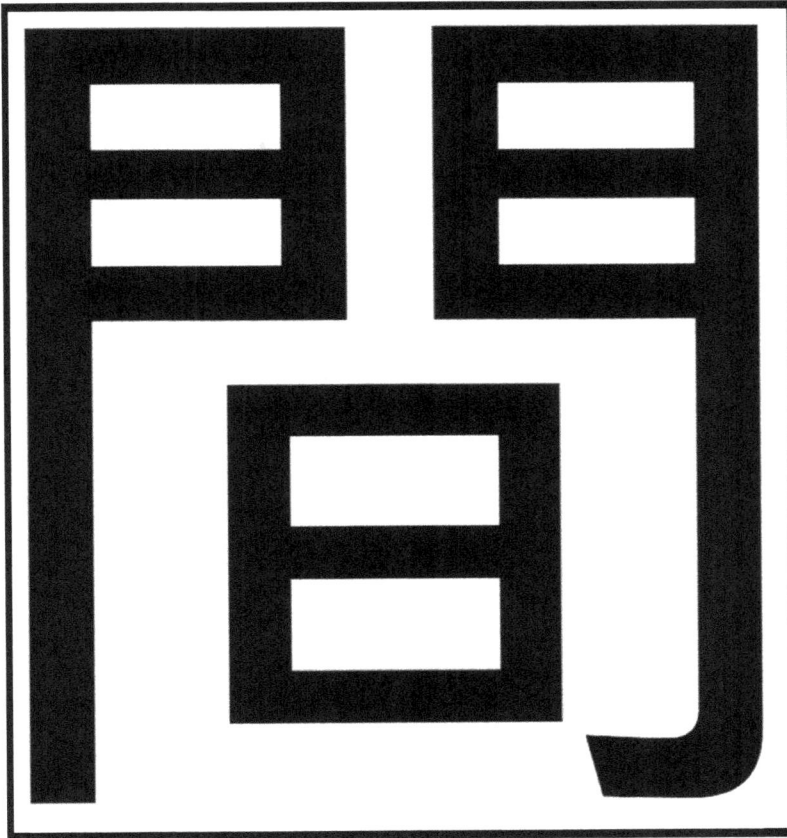

Figure 10.4 – Japanese word 'Ma'

In music, it is the silence between the notes that gives structure to the piece. In the practice of meditation, there is a point of being conscious of breathing and the pauses between breaths. The awareness of the gaps between inhaling and exhaling bring clarity to thought and allows for the observation of stillness or turbulence of thought. So how do art, music, meditation and balanced design relate to engineering of complex systems?

As environments grow and evolve over time, the number of services, layers and systems grow as well. The systems are no longer represented accurately by simply referring to them as the services or the uptime metric. They have become bigger than that, and the interaction between the components is what needs to be monitored. If you were to look at it from a design perspective, this is what would be considered the whitespace. It is the thing that is in-between "here" and "there", but it is often ignored because you can't see it. It is ephemeral and shapeless and can only be seen when it is absent. It is arguably the most important aspect of design and architecture.

Distributed software systems are evolving at an ever-increasing rate. The modular nature of them allows for continuous development and code commits. How can assurances be made that the interactions between all components will always behave in a consistent and expected manner? What happens when some of these components don't play by the rules? What if they stop communicating, generate bogus data, or outright fail? Software is designed with the assumption that it will work, so how do organizations plan for the fallout of when it does not? What gets affected in the cascade of failure when services have a hissy fit and wig-out? Or to put it in other terms, to have an escalating digital tantrum and operate out of the realm of expectation.

The old school of thought was to:

1. Minimize downtime by ensuring services are monitored, as are latencies and user interaction response time.
2. Have a NOC (Network Operations Center) to monitor the statistics and then respond quickly to incidents, troubleshoot, resolve, then do a root cause and corrective action (RCCA).

This doesn't work for critical services anymore, because the risk of downtime becomes more likely as complexity increases. Complexity also increases the time to resolution due to the increase in places where failures may occur. The industry as a whole was grappling with this problem in the last few decades until a solution started to reveal itself from an unlikely source.

In 2001, a firefighter and EMT for Cirque du Soleil by the name of Jesse Robbins was changing jobs and applied for two at the same time. One was a bus driver, and the other was a backup engineer at Amazon. Amazon called back first and Jesse started working for them. He brought his mindset and experience in emergency services to the table and transformed not only the way Amazon approached critical services, but the entire industry.

Jesse founded the ideology of Chaos Engineering and was titled with the role of *Master of Disaster*. His approach was to proactively fail systems to examine what breaks and fix it in a controlled environment before it happens in the real world. It was iterative and highly controlled, like an ad-hoc DR test for every service and system in the environment. After enough iteration, entire datacenters could be taken offline and customers would not experience any issues.

You can think of this in firefighting terms as a *controlled burn*. When a forest fire occurs, one technique that is used to stop the spread of the fire is to remove the fuel. This is done by proactively burning areas in smaller, controllable sections and digging out other areas, making sure that there is a dead zone where there is no fuel for the forest fire to use and to continue spreading. By being proactive, and having a small burn, you prevent the big one.

In 2006, Jesse left Amazon and started socializing his methods via blogs and social media. He then created the conference named Velocity, which was used by massive organizations like Microsoft and Google to discuss solutions to infrastructure problems.

Other massive organizations started to use the methodologies in their own way. Yahoo created a system called Nemesis that allows for controlled fault injection. Netflix created a platform called FIT (Fault Injection Testing), which allows them to have controlled outage of many aspects of their environment.

Figure 10.5 – Netflix simian army

Also of Netflix fame, is the Simian Army, which started with a tool called the Chaos Monkey. The Chaos Monkey would randomly kill services across the enterprise, allowing departments to solve dependency issues and ensure that any single failures that may occur in the future, do not affect the availability of services. They would bolster some areas and isolate others. To protect against a single point of failure (SPOF), service-centric clusters were created across many nodes that could spawn quickly to scale-up or terminate to scale-down. Nodes in those clusters could also terminate and re-spawn quickly if there is any hint of a fault. This was fantastic for the design of high availability of services, but other areas beyond availability could benefit from chaos engineering as well. Thus, Netflix created the Simian Army.

Other members of the Simian Army include:

Conformity Monkey will identify servers, (known as instances in Amazon) that do not adhere to defined best practices and then it will terminate them.

Security Monkey will identify instances, which have security or vulnerability violations and then terminate them.

Doctor Monkey looks at resource utilization like memory and CPU, and then will remove unhealthy instances from service. The service owners then determine what the cause was and respawn the instance.

Latency Monkey can simulate degradation of service response and determine what dependencies are affected. This allows for failover testing without taking an instance down.

Chaos Gorilla induces the outage of an Amazon availability zone. This is useful for testing global load balancing strategies.

10-18 Monkey will look for regional instances whose config does not adhere to the localization configuration. Things like languages and character sets are evaluated.

Janitor Monkey looks for unused resources and terminates them.

Chaos Kong will take down a region, which is comprised of multiple availability zones. This one is serious and is not employed lightly.

A commonality that you will find in many of the chaos engineering tools from various organizations is that they are strictly for Amazon Web Services. However, some projects transcend AWS and work with other platforms or even multiple environments. A few of them are:

Chaos Gopher (generic framework)

Chaos Lemur (Pivotal Cloud Foundry – AWS, VMware, Openstack)

WAZMonkey (Microsoft Azure)

Some key points that have been socialized by the teams that created the chaos engineering tools to minimize the risk of unintended consequences are these:

1. Define a "steady state", or "baseline" for all services and systems. When a chaos tool is deployed, look for variances from the baseline. Then dig into the relationship between the chaos tool and the issue.
2. When a chaos tool is deployed, the goal is to ensure that there is no user impact during the running of said tool. If there is an impact, observe the cause and effect, then fix the problem and run again. After enough iteration of fixes, the environment should be resilient to anything that the chaos tool can throw at it.
3. When a chaos tool is initially deployed, a small set of services should be included in the scope to possibly be affected. Once the tool has been running for a few months and there are no longer any adverse effects to the user experience (of the affected service) during a chaos event, then add additional services can be added to the scope.
4. Once most services are added to the chaos tool and the iterative fixes have been applied, then any new chaos tools that are added will have all services in their scope automatically unless explicitly defined in an opt-out list.
5. Service owners should make all efforts to make their services resilient enough to be removed from the opt-out list. Opt-out lists would be reviewed in detail once a month.
6. During a real customer impacting event, all chaos tool activities should be halted until the issue is resolved.
7. Have a blameless culture in RCCA's and post-mortems. Enterprises will be stronger by going through the process, but there will be bumps and outages along the way.
8. Take the learning from the chaos engineering process and modify core architecture designs and recommended practices going forward.
9. Ensure that the chaos tools run continuously and via automation in production.

10.7 Chapter Summary

1. Red Teams provide insight into the security holes that are not so apparent and would not likely be found by standard vulnerability assessments.
2. Red Teams can be used as a tool to justify a security budget to senior management.
3. Threat profiles determine the intensity, stealth, duration, number of attackers and their capabilities. No organization can withstand a level 1 threat.
4. Security maturity is necessary for a Red Team to be truly effective. Several iterations of pen tests and vulnerability assessments must have occurred beforehand to ensure the Red Team engagement is worthwhile.
5. An infrastructure will be in one of 4 states: unpatched, managed, in-flight and road mapped.
6. Chaos engineering induces faults so they can be fixed in a controlled environment.
7. Baselines must be performed before chaos tools are deployed and incredibly detailed monitoring is a pre-requisite.
8. Chaos tools should be introduced in a staged manner to ensure the least amount of user impact.

10.8 Chapter Review Questions

1. Classify these threat profiles:

    a) A script kiddie performing attacks directly from their computer at home.
    b) An advanced persistent threat (APT) on a retail store chain by an organized cybercrime syndicate.
    c) A state funded international cyber-terrorism campaign with a dedicated wing of a military group and a political agenda.
    d) A disgruntled employee with full admin level access who has worked there for 10 years and is getting fired without just cause.

2. In what order would a Red Team attempt to perform the following and why?

    a) Physical security penetration.
    b) Social engineering.
    c) External network access.
    d) Internal network access.

3. What does chaos engineering force an organization to do?

4. What is a recommended practice for chaos tools to be doing when there is no customer impacting issue occurring? What about when there is?

CHAPTER 11

Frameworks and Certifications

"I cannot teach anybody anything. I can only make them think."
--Socrates (Philosopher)

If you want to go farther down the rabbit hole and get serious about formal risk management practice, then this is the chapter for you. Here you will compare various risk management frameworks and see a possible certification path, depending on your career goals.

11.1 The ISACA Risk IT Framework

There are several paths that you can follow if you want to continue learning how to understand and articulate risk. The most notable is by following the ISACA Risk IT Framework. The framework covers the areas of risk governance, risk evaluation and risk response.

ISACA (formerly named the Information Systems Audit and Control Association) is an association that was formed in 1967 to centralize the informational guidance for organizations that had computer systems as a critical part of their operations. The Risk IT Framework was first published by ISACA in 2009 as a result of collaboration from industry experts and academics from around the world.

11.1.1 Risk Governance deals with the topics of

- Risk Culture
- Communication and Awareness
- Risk Tolerance
- Risk Appetite

It is Risk Governance that identifies the risk exposure of the organization and defines the controls that will be adhered to, and measured against, to reduce identified risks. It requires a very precise and patient personality to ensure that business processes are followed or modified. Many discussions with senior management are required.

11.1.2 Risk Evaluation deals with the topics of

- Creating Risk Scenarios
- Creating Business Impact Analysis Reports
- Determining Probability and Criticality

When conducting tasks associated with Risk Evaluation, an extremely well-rounded skillset is needed. Many conversations with stakeholders are required to get a solid understanding of the entire physical and logical infrastructure, business processes and workflows. An understanding of infrastructure history and the future technology plan of the organization is also required so that past risk responses and future exposures can also be accounted for and analyzed.

11.1.3 Risk Response deals with the topics of

- Identifying key risk indicators (KRI)
- Developing possible responses to the identified risks
- Prioritizing risk responses based on criticality, cost and difficulty of remediation

When performing risk response related tasks, there is a lot of project management and Gantt charts involved. The only way to ensure that the responses are performed as required is to get very granular with the task definitions and to follow through with their remediation. Planning with multiple teams, management and operations is needed. If buy-in on the remediation is not universal across all stakeholders, then the response actions will stall or fail. It is the role of the person managing the risk response to ensure that everyone is on the same page and accepts the strategy and timelines.

11.2 Other related frameworks

There are other frameworks that can also be considered for use and are provided in this section.

11.2.1 COSO Enterprise Risk Management Integrated Framework

COSO (or the Committee of Sponsoring Organizations of the Treadway Commission) was created in 1985 to address corporate fraud. The organization created a framework for management, business ethics, organizational governance and enterprise risk management.

In 2004, COSO released the Enterprise Risk Management Integrated Framework. It is comprised of eight framework components, listed below:

- Internal environment
- Objective setting
- Event identification
- Risk assessment
- Risk response
- Control activities
- Information and communication
- Monitoring

While the COSO frameworks can be adapted, and applied to any organization, they are used most frequently in large organizations that conduct themselves internationally. Often, there are internal auditors that monitor compliance as well as external auditors that verify it. COSO has published some guidance for smaller companies, but it is limited in scope to internal controls and financial reporting.

11.2.2 VAL IT Framework

The VAL IT framework, also created by ISACA, provides guidance on getting the most value from your IT investments. It deals with topics such as optimization of lifecycles, resource allocation, mapping ROI to IT investment and other processes related to maximizing business value. It has now been fully integrated into COBIT.

11.2.3 COBIT – Control Objectives for Information and Related Technology

COBIT is used to define processes for day to day operations for IT. It aligns controls with the business objectives of the organization, makes them measurable and provides accountability.

COBIT was first published by ISACA in 1996 with a limited scope around auditing. Since then, there have been 4 subsequent versions that incorporate more areas including; controls, management and IT governance. As of COBIT version 5, published in 2012, it also encompasses the RISK IT Framework and the Val IT framework.

11.2.4 ISO 31000

The ISO 31000 family of standards includes the following:

- ISO 31000:2009 – Principles and Guidelines on Implementation.
- ISO 31010:2009 – Risk management & risk assessment techniques.
- ISO guide 73:2009 – Risk management vocabulary.

It provides standards and best practices for risk management that can apply to all industries. It is often used to homogenize and improve existing risk management practices within an organization that have been created organically.

11.3 CRISC Certification

CRISC (pronounced see-risk) stands for Certified in Risk and Information Systems Control. It is the only well-known risk related certification out there. This certification, created by ISACA, requires a written exam and a minimum of three years' worth of experience in a minimum of two CRISC domains. This experience can be gained either before or after the exam. However, the CRISC designation will not be granted until enough experience is gained and proof is supplied in an application to ISACA.

The CRISC domains are as follows:

1. Risk identification.
2. Risk assessment.
3. Risk response and mitigation.
4. Risk and control monitoring and reporting.

There are no prerequisites for taking the exam. The experience requirement needs to be verified with employers by ISACA. If the experience requirement is not met within 5 years, then the exam must be retaken to the achieve certification.

The recommended preparation for the CRISC certification is:

1. Obtain and study the CRISC review manual.
2. Obtain and study the CRISC review questions, answers and explanations manual.
3. Obtain access to and review the CRISC practice question database.
4. Perform related work in two or more CRISC domains.

11.4 Actuarial Sciences for IT

An actuary is a mix of an accountant, a business analyst and a statistician. They can determine the probability of future events based on data from many sources. An actuary can determine the financial impact of identified risks and they can help organizations formulate appropriate responses.

Most actuaries work within the financial sector, primarily in the realm of insurance. In enterprise organizations, they often have the role of Chief Risk Officer acting to define enterprise risk management and governance programs.

Actuaries are not normally known to operate solely within the realm of IT, but many of the concepts, tenets and goals of actuarial sciences are in concert with the objectives of IT. That is, the alignment of technical and operational strategies within business objectives.

One of the goals of this book is to allow IT practitioners, regardless of position or role, to have a greater understanding of the many moving parts of an organization and how they can influence the design and operation of an IT infrastructure.

By following the same principles as actuaries, and looking at the big picture, the business requirements, the processes, the culture, the detailed history of an organization and its planned future growth, you can develop detailed future models and plan with greater accuracy.

Unfortunately, there is no one certification for this. There is no one well defined path or designation that you end up with in the end. For every person, the result will be different, but as a community we can work to elevate the body of knowledge from which we work and the standards that are set in the industry. There could be more informed Administrator and Architects who understand how risks relate to business and technology decisions. Management could understand the infrastructure and how a design directly relates to business objectives. There could be more communication and collaboration across departments, removing silos and strengthening the organization.

There are many methods to achieve this, but the first step is introspection. Look at your own organization and start applying some of the techniques, processes and thought patterns presented in this book. I hope the consideration of, and detailed responses to, identifiable risks to an organization's infrastructure has allowed you to see things in a new light.

I encourage readers to also continue on with the IT Architect Series. The next book in the series "The Journey", details the soft skills and techniques one can use to advance their career. If you want to reach a level of qualification that is considered "an expert", then it is definitely a book that will help. If this book is your first exposure to the series, then you must also read the first book "Foundation in the Art of Infrastructure Design". It is a body of knowledge that is indispensable.

11.5 Chapter Summary

1. The ISACA Risk IT Framework is comprised of risk governance, risk evaluation and risk response.
2. COBIT 5 encompasses the ISACA Risk IT Framework.
3. The ISO 31000 family of standards is often used to clean up and homogenize organically created Risk Management practices.
4. CRISC is the only well-known risk specific certification in the industry.

11.6 Chapter Review Questions

1. Review the components of the ISACA Risk IT Framework and state the following:

   a) Which component of the framework deals with identifying KRIs?

   b) Which component of the framework requires the most communication with senior management?

2. How do you achieve CRISC certification without work experience within the CRISC domains?

Appendix A - Glossary of terms

3-way handshake

A method of verifying packet transmission in TCP SYN, SYN \ ACK, ACK.

AAA

Access, Authentication, Accounting.

Acceptable risk

The amount of risk that an organization is willing to tolerate.

Active redundancy

Active redundancy utilizes an embedded monitoring and response system that logically removes failed hardware.

Actors

Parties that are involved in an interaction, often negative. Example: malicious actors.

Actuary

An actuary is a mix of an accountant, a business analyst and a statistician.

Agent Theory

In agency theory, there are two parties; the principle and the agent. The agent is able to make decisions on behalf of, or that affect, the principle. If the two parties have different interests, and the information flow is not pervasive, then the agent may act selfishly at the expense of the principle.

Agile IT

The use of Agile processes allows for multiple concurrent projects to occur with rapid iteration and incremental gains.

Application Clustering

Application clustering provides a mechanism of availability that has a dedicated heartbeat between servers. If the heartbeat goes down, then the surviving server will take over the identity of the cluster and continue serving data.

Assumptions

The unknown facts that are estimated in order to complete a hypothesis or a solution.

Availability

The measure of ability to continue to provide services.

Availability Zones

An availability zone is a data center within a region.

BC/DR

Business Continuity / Disaster Recovery.

Bias

Prejudice in favor or against something.

Black Swan Event

A dramatic event that does not occur very often, is difficult to predict and has far reaching effects.

Blue Team

A team of people that provide defensive capabilities against a Red Team.

BRD

Business Requirement Document.

Business Continuity

The ability for an organization to continue business operations in the face of a failure.

BYOD

Bring Your Own Device.

CAFTA

Computer Aided Fault Tree Analysis.

Carelessness

The personal trait of not caring, or putting little effort into accuracy or quality of work.

CDN

Content Delivery Network.

Chaos Engineering

A method of proactively failing systems to examine what breaks and then fix it in a controlled environment.

Chaos Theory

A branch of mathematics that focuses on the variances created in dynamic systems due to sensitive initial conditions.

Cheap

When cost is the prime factor in every decision, despite consequences.

CI/CD

Continuous Integration / Continuous Delivery.

Cloud Native Applications

Applications that are designed for elasticity using microservices.

Cluster Witness

An application that monitors cluster nodes and assists in the decision making of which side remains active after a loss of communication between nodes.

CMDB

Configuration Management Database.

COBIT

Control Objectives for Information and related Technologies.

Cognitive Bias

An error in reasoning that is usually associated with preconceived notions or preferences.

Conceptual Design

An early phase of design that broadly articulates the intention with simplistic models.

Conscious Incompetence

When someone is unable to do something and they are aware of it.

Constraints

A limitation or restriction that is inherent to a situation.

Consumer-Grade

Reflecting product quality that is mass produced and focused on lower cost, as opposed to a higher quality of engineering.

Content Delivery Network

A system of distributed servers that provide content to users at a high speed, based on their region This ensures that the closest server provides the content.

Contingency

A provision for responding to possible future events.

Continuous Delivery

A software engineering method where teams produce software in shorter cycles.

Continuous Integration

A software development practice that requires developers to integrate code changes to a shared repository very frequently.

COSO

Committee of Sponsoring Organizations.

CRISC

Certified in Risk and Information System Control.

Culpability

The blame or fault of an incident or situation.

Data Locality

The physical location where data resides.

Data Redundancy

Having multiple copies of data in different locations.

Decision Theory

The theory that every decision can be categorized into three categories: rational, non-rational and irrational.

Decisions with Certainty

These are logic based decisions that have all variables defined and quantified.

Decisions with Risk

These are decisions with some level of probability that will affect the outcome.

Decisions with Uncertainty

These are decisions with unknown-unknowns.

Defense in Depth

A strategy to have multiple layers and techniques of defense.

Demilitarized Zone

In networking, it is an isolated zone that has some public access, but is cordoned off from the rest of the internal network.

Dempster-Shafer Theory of Belief

A mathematical method for determining a quantifiable level of trust in a hypothesis.

Dependencies

When there is a reliance between two or more people or things. Such as applications services relying on database services to function.

Design Decisions

Decisions that are made during the architecture design process. They are an output from analyzing constraints, assumptions, risks, requirements and dependencies.

DevOps

A movement in IT that automates operations by treating infrastructure as code.

Disaster Avoidance

Disaster avoidance is the process of reducing the probability of a disaster and minimizing the impact if one does occur.

Disaster Recovery

The strategy that an organization employs to respond to a disaster in order to ensure that critical business functions become available as quickly as possible.

Disgruntled

Angry or dissatisfied.

DMZ

Demilitarized Zone

Elastic Infrastructure

Infrastructure that can adapt to demand by provisioning or de-provisioning resources.

Enterprise Risk Management

The process of planning, organizing, leading, and controlling the activities of an organization in order to minimize the effects of risk.

ERM

Enterprise Risk Management

Event Tree Analysis

A modeling technique that explores responses originating from initiating event and maps out the probabilities of the outcomes for analysis.

Exfiltrate

To surreptitiously remove something.

Fail-Fast

A fail-fast system will proactively fail and be removed from production when it does not perform within the desired thresholds.

Failure Mode Effect Analysis (FMEA)

A systematic technique for failure analysis.

Fault Tree Analysis (FTA)

A method of failure analysis using Boolean logic.

Five Nines

A very high level of uptime. 99.999% available.

Framing

When people respond differently to options, based on how they are presented.

FRD

Functional Requirement Document.

Graceful Degradation

The ability to retain some functionality after one or more failures.

Guerilla

Unstructured. Non-traditional.

Heuristics

A problem-solving method that provides a quick and workable approximate solution. Often used in artificial intelligence and machine learning.

High Availability

A measure of uptime for infrastructure of services.

Host Based Clustering

When resources, such as memory, cpu, storage, are put into a resource pool.

Human Error

"This sort of thing has cropped up before, and it has always been due to human error." – Words of wisdom from HAL (2001, A Space Odyssey).

HVAC

Heating Ventilation Air Conditioning.

Hypochondria

Unwarranted fear and anxiety about health.

IDS

Intrusion Detection System.

Infrastructure Diversity

Utilizing technologies from many vendors to diversify the risk associated a single one that may cause a cascading failure.

Inherent Risk

The risks before controls are in place.

IOC

Indicators of Compromise.

IPS

Intrusion Protection System.

ISACA

Information Systems Audit and Control Association.

ISO

International Standards Organization.

IT Acumen

Skill in IT gained through experience.

IT agility

The ability to adapt quickly to changing requirements in the realm of IT.

Kaplan-Meier curve

A technique for statistical estimation.

KRI

Key Risk Indicators.

LD50

The point at which 50% of test subjects die from a particular test.

LMS

Log management system.

Load Balancing

Distributing workload across multiple resources.

MWI

Many Worlds Interpretation.

MART

Mean Active Repair Time.

MFDT

Mean Fault Detection Time.

Microservices

A means by which large services are built on smaller modular services that work together in a scalable manner.

MRT

Mean Repair Time.

MTBF

Mean Time Between Failures.

MTPoD

Maximum Tolerable Period of Disruption.

MTTF

Mean Time to Failure.

MTTR

Mean Time to Repair.

MWI

Many Worlds Interpretation.

Myers-Briggs Type Indicator

A means of categorizing personalities into 16 discrete types.

NFRD

Non-functional Requirements Document.

NMS

Network Monitoring System.

Off-The-Shelf (OTS)

Easily purchased through a store, with no rework or further integration required.

On-prem (On-premises)

Physically in a building belonging to the organization using it.

Open Source

Software that has its source code freely available and may be redistributed and modified.

Opportunity Cost

The loss of a potential gain when another option is chosen.

OSSM

On-demand, Self-service, Scalable, Measureable.

P-F Interval

The amount of time between a Potential Failure and a Functional failure.

Passive Redundancy

When a device has redundancy that activates when a failure occurs.

PDU

Power Distribution Unit.

Pen Testing

Penetration Testing.

Physical Design

A very detailed technical design that can be used as a build guide.

Pretext

A justification for a course of action that hides the real motive.

Probability

The extent that something is probable.

Procurement

The act of obtaining something.

PXE

Pre-eXecution Environment.

Qualitative

Measuring the quality of something rather than quantity.

Quantitative

Measuring the quantity of something rather than quality.

Red Team

The practice of viewing a problem from the perspective of a hostile actor.

Residual Risk

The amount of risk that remains after a risk response is taken.

Resiliency

The ability to withstand stressors.

Resilient Computing

The ability to maintain an acceptable level of service despite difficulties and faults.

Rework Maintenance

When a component is removed from production, but still has usefulness or value if repaired or modified, then it is reworked.

Rework Validation

When a component is modified in the rework process, it needs to be validated before it can be used with certainty.

RISC

Reduced Instruction Set Computer.

Risk

A situation that exposes danger.

Risk Accountability

The person(s) ultimately responsible for addressing a risk.

Risk Appetite

The amount of risk that an organization is willing to accept.

Risk Awareness

The understanding of what risks exist.

Risk Factor

Things that increase or decrease the impact or frequency of a risk scenario.

Risk Governance

The rules and mechanisms that an organization employs as risk management strategies.

Risk Impact

The consequences if risk events are realized.

Risk Management

The identification, assessment, and prioritization of risks.

Risk Model

A method of representing an all-encompassing view of the effects that may occur from a risk event.

Risk Register

An inventory of all known risks within an organization.

Risk Response

The action taken in response to a risk.

Risk Scenario

A proposed situation that provides the basis for one or more risk event to occur.

Risk Transparency

The level at which risks are socialized within the organization.

Risk Treatment

The development of risk responses.

Risk Vector

Risk vectors are the means in which risk can manifest as an event in an environment.

Root Cause Analysis (RCA)

A method of problem solving that identifies the root cause of faults.

Root Cause and Corrective Action (RCCA)

A method of problem solving that identifies the root cause of faults that have or may occur and provides possible corrective actions.

Rubicon

A river in northern Italy famously crossed by Julius Caesar in 49BC.

SaaS

Software as a Service.

Schrödinger's cat

A cat imagined as being enclosed in a box with a radioactive source acting as a poison that may be released. When the source emits radiation, the cat is considered to be simultaneously both dead and alive until the box is opened and the cat observed.

Self-healing Systems

A method to which software can identify a fault and automatically apply corrective action to return it to a desired state.

Service Level Agreement (SLA)

A contract between a service provider (either internal or external) and the end user that defines the level of service expected.

Service Level Objective (SLO)

A single performance characteristic that is a component that comprises an SLA.

Single Loss Expectancy

The monetary value associated with the occurrence of a risk event against a single asset.

Social Engineering

The use of deception to convince individuals into divulging information that may be used for fraudulent purposes.

Survivability

The capability of withstanding a risk event.

Survivability Analysis

Survival analysis is a branch of statistics for analyzing the expected duration of time until one or more events happen that may cause a functional failure.

Taxonomy

A means of classification.

TCO

Total Cost of Ownership.

Tetrimap

The creation of a treemap by use of shapes similar to the ones used in the game "Tetris".

Tetris

A Russian tile-matching puzzle game.

Threat profile

The profile of a malicious actor that indicates the sophistication or vigor in which they can attack the environment.

Tribal Knowledge

Knowledge that is passed down orally and is not written. It is well guarded and only provided to those that are deemed worthy.

Unconscious Incompetence

When someone does not realize that they do not have the skills to perform a task.

Unicorn Zone

A theoretical zone where things can be done extremely quickly, very inexpensively and the outcome is of high quality.

UPS

Uninterruptable Power Supply.

Vulnerability Assessments (VA)

An assessment done on an organization that indicates missing patches and known vulnerabilities.

VAL IT Framework

It provides guidance on getting the most value from your IT investments.

WAF

Web Application Firewall.

www.ingramcontent.com/pod-product-compliance
Lightning Source LLC
Chambersburg PA
CBHW082127210326
41599CB00031B/5895